SpringerBriefs in Applied Sciences and Technology

PoliMI SpringerBriefs

For further volumes:
http://www.springer.com/series/11159
http://www.polimi.it

Salvatore Carlucci

Thermal Comfort
Assessment of Buildings

POLITECNICO
DI MILANO

Springer

Salvatore Carlucci
Energy Department
Politecnico di Milano
Milan
Italy

ISSN 2282-2577   ISSN 2282-2585 (electronic)
ISBN 978-88-470-5237-6 ISBN 978-88-470-5238-3 (eBook)
DOI 10.1007/978-88-470-5238-3
Springer Milan Heidelberg New York Dordrecht London

Library of Congress Control Number: 2013930608

Printed on acid-free paper

Springer is part of Springer Science+Business Media (www.springer.com)

Foreword

The wealth of research on thermal comfort has been partially taken and crystallized into international standards, where thermal comfort is defined as: "that condition of mind which expresses satisfaction with the thermal environment and is assessed by subjective evaluation". A selection of subjective judgment scales has been described, e.g., in ISO 10551. Those scales propose a set of answers to questions as: "how do you feel at this precise moment?", or "please state how you would prefer to be now", so they allow collecting information about the thermal sensation and preference of a certain subject in a given place at a given time.

The data collected via these standardized surveys in the laboratory and in the field have been interpreted, and meaningful correlations between the answers and various physical variables have been derived, giving rise to what are generally called comfort models, for example, the Fanger whole-body steady-state heat balance model, the Pierce two-node model, the adaptive models and others. All these models have as input the *here-and-now* questions and make *here-and-now* predictions over the likely answers of a group of people in a certain environment.

But, when assessing comfort performances of an existing building or using a certain comfort target interval as one of the objectives of a building design process, one is generally interested in the overall performance. So one would attempt to consider some adequate average over time (e.g., a season, a year, etc.) and space (e.g., all occupied thermal zones of a building) of the *here-and-now* thermal comfort values, be them gathered via direct interviews in a building or calculated via one of the models. Disparate averaging algorithms have been proposed in the literature, and some are presented in the standards and available for use in applications.

All this at least in theory; in everyday practice budget constraints and other limitations have often led to using very simplified rules for assessment or design, even not making explicit which model and assumptions are taken as a basis. Averaging algorithms have been used often without an analysis of their implications on design choices, and very limited comparison between them has been performed.

But in the last years, under the renewed effort toward low- and zero-energy buildings, the issues of fine-tuning comfort and fully understanding its connection with energy use have become increasingly important and urgent to address, particularly so in warm climates and warm periods.

A number of European research projects (e.g., SCATs, Commoncense, ThermCo, KeepCool) have explored these issues and added new data to the comfort databases about occupied real buildings; conferences and networks such as NCEUB, Palenc, and IEA SHC Task 40/ECBCS Annex 52 have been a fruitful research cooperation and exchange opportunity for analyzing the implications on comfort design; some of the new findings have found their way to the recent update of the standards EN 15251, ISO 7730, and ASHRAE 55, and will influence their further ongoing revision.

The research work of Dr. Carlucci presented in this book represents an important contribution to these advancements and a fruit of his active engagement in some of the mentioned projects and networks, in the framework of his participation in the end-use Efficiency Research Group of Politecnico di Milano.

A careful review, comparison, and analysis of the large number of long-term indexes proposed in the literature were highly needed and are now hence available. Building on those, Carlucci proposes a new improved long-term general discomfort index which aims at better matching the specific objectives of real world assessment and design and to be applicable with the three main comfort models presented in the standards. It also explicitly defines the operational use of the index (e.g., how to define the length of the calculation period based on the actual climate of the site) in order to overcome the present ambiguities that often undermine the gnoseological and practical relevance of the results. Finally, he developed three computer codes in the EnergyPlus Reference Language for calculating the three versions of the new index and integrated them in the simulation environment EnergyPlus in order to calculate the new index and to report it as a direct output of the simulation.

Overall, a clear-cut methodology is here an essential tool to produce useful results for real world applications.

Lorenzo Pagliano
Politecnico di Milano

Acknowledgments

I wish to express my sincere gratitude to Prof. Edward Arens and Prof. Matheos Santamouris for their valuable suggestions. I thank warmly Prof. Lorenzo Pagliano for having supported me during the execution of this work and for having reviewed it. I wish to express my gratitude also to Prof. Gabriele Masera for the inspiring discussions that helped me in developing the topic. I am also grateful to my colleagues and friends Dr. Paolo Zangheri, Eng. Marco Pietrobon, and Francesco De Rosa who helped me in times of need. Infinite thanks to my family, whose constant support helped me to complete this work. And of course, thanks to you, Natascia, for your support and your even greater patience.

Salvatore Carlucci

Contents

Symbols and Abbreviations

Symbols

α	Solar absorbance of a surface (dimensionless)
B	Digit binary code: 0–1 (dimensionless)
γ	Solar factor (%)
h	Heat transfer coefficient (W m^{-2} K^{-1}) or Hour (h)
I	Global solar irradiance on a horizontal surface (W m^{-2})
v_a	Average air velocity (m s^{-1})
PMV	Predicted mean vote (dimensionless)
PPD	Predicted percentage of dissatisfied (%)
U	Steady-state transmittance (W m^{-2} K^{-1})
θ_{op}	Operative temperature (°C)
θ_{db}	Dry-bulb temperature (°C)
θ_{mr}	Mean radiant temperature (°C)
θ_{op}	Operative temperature (°C)
θ_{os}	Sol-Air temperature (°C)
θ_{res}	Dry-resultant temperature (°C)
θ_{rm}	Running mean of outside dry-bulb temperature (°C)
wf	Weighting factor (dimensionless)

Subscripts

actual	Actual status
actual PMV	Referred to PMV calculated in actual status
c	Convective
C	Cold period
comf	Comfort
$\bar{d}$	Value averaged on a day
lower limit	Lower limit of comfort range
OC	Overcooling
OH	Overheating
out	Outdoor

PMV limit	Referred to PMV limits
i	General recursive index
in	Indoor
r	Radiative
t	Recursive index for time
upper limit	Upper limit of a comfort range
W	Warm period
Y	Year
z	Recursive index for zones
Z	Number of zones in a multi-zone building

Acronyms

ANSI	American National Standards Institute
ASHRAE	American Society of Heating, Refrigerating and Air Conditioning Engineers
CEN	European Committee for Standardization
CIBSE	Chartered Institution of Building Services Engineers
Dh	Degree-hours
DhC	Degree-hour criterion
US DOE	Unites States Department of Energy
DSY	Design Summer Year
ECBCS	Energy Conservation in Buildings and Community Systems
EMS	Energy Management System
EN	European Standards
ERL	EnergyPlus Runtime Language
EU	European Union
HVAC	Heating, Ventilation and Air Conditioning
IEA	International Energy Agency
ISO	International Organization for Standardization
IWEC	International Weather for Energy Calculations
LPD	Long-term Percentage of Dissatisfied
NaOR	Nicol et al.,'s Overheating Risk
NREL	National Energy Renewable Laboratory
PMV	Predicted Mean Vote
POR	Percentage Outside (comfort) Range
PPD	Predicted Percentage of Dissatisfied
PPDwC	PPD-weighted criterion
RHOR	Robinson's and Haldi's Overheating risk
SCATs	EU Project Smart Controls and Thermal Comfort
SHC	Solar Heating and Cooling Program
SIA	Swiss Society of Engineers and Architects
Sum_PPD	Accumulated PPD
TMY	Typical Meteorological Year

TRY	Typical reference Year
TRNSYS	Transient system simulation program
USA	United States of America
WYEC	Weather Year for Energy Calculations

Introduction

The specification of indoor thermal comfort requirements that a building must provide is a prerequisite for its design, and reliable explicit methods for the assessment of its long-term comfort performances are, therefore, necessary. Several metrics for assessing human thermal response to climatic conditions or stresses have been proposed in the scientific literature over the last decades, and a number of authors have used, and still use, terms such as *discomfort index, stress index,* or *heat index* to identify the analytical models that describe human thermal perception of the thermal environment to which an individual or a group of people is exposed. More recently, a new type of *discomfort index* has been proposed in the scientific literature, in standards and guidelines, specifically for briefly describing long-term thermal comfort conditions in buildings and for predicting uncomfortable phenomena, in particular summer overheating. Most of these new indices summarize the thermal performance of a building into a single value.

These indices may be useful tools for the operational assessment of the thermal comfort performance of an existing building or for guiding the optimization phase of the design of a building envelope and its thermal plant systems and control strategies. In particular, for zero energy—mainly passive—buildings, the possibility to discriminate and rank building variants is not satisfactorily feasible by comparing the energy consumption (ideally the best variants will all have energy consumption values grouped in a small interval around zero). We argue here that the ranking of these variants may be explicitly based on maximizing thermal comfort performances of the envelope, passive systems, and their control strategies. This also coincides with minimizing energy need (and hence energy consumption of active systems whether present) for achieving comfort design values, but is more flexible.

In Chap. 1, a hopefully exhaustive review of the existing indices for the long-term evaluation of thermal comfort conditions in a building and for thermal risk assessment is presented (i) since some of the them are based on thermal comfort models, while others derive from rules of thumb, (ii) since they are considerably different in their structure and significance and (iii) since a systematic collection of those is missing.

In Chap. 2, 16 long-term discomfort indices were collected, compared, and contrasted on the ranking of 54 different building variants obtained by varying four key parameters of the design of a large reference office building characterized by 34 zones. Similarities and differences in their ranking capability are derived from this analysis.

In Chap. 3, a gap analysis is performed in order to assess their dependency on specified boundary conditions. Since the dependence on certain boundary conditions is sensitive, a calculation framework is proposed in order to suggest a homogeneous and harmonized structure for their calculation.

Following the previous analysis, a new improved long-term general discomfort index is proposed in Chap. 4. It is called Long-term Percentage of Dissatisfied (LPD) and it has been designed to be used with the three comfort models currently in force; one version is based on the Fanger comfort model, another on the European adaptive comfort model, and the last on the American adaptive comfort model. In order to effectively assess the likelihood of dissatisfied with the American adaptive comfort model, an analytical formulation of the relationship between the offset in degree of the actual indoor operative temperature with respect to the optimal comfort temperature and the likelihood of dissatisfied has been derived from a logistic regression analysis of the data collected in the ASHRAE RP-884 database. This mathematical relationship has been called ASHRAE Likelihood of Dissatisfied (*ALD*).

Moreover, three computer codes have been written in the *EnergyPlus Reference Language* for calculating the three versions of the new index and have been integrated in an energy dynamic simulation environment (EnergyPlus) in order to calculate the new index and to report it as a direct output of the simulation.

The author is currently contributing to the ongoing work for analyzing and clarifying a possible definition framework for net zero energy buildings within the International joint project IEA SHC Task 40/ECBCS Annex 52, with the explicit inclusion of quantitatively defined thermal comfort objectives as an integral part of the definition of a zero energy building.

Chapter 1
A Review of Long-Term Discomfort Indices

Abstract Over the past decades, several metrics and methods have been developed to assess human perception of the thermal environment and thermal response to different climatic conditions. In the last decade, a new type of discomfort index, intended to describe long-term thermal phenomena in buildings has been proposed in the scientific literature, standards and guidelines. Most of these new indices summarize the thermal performance of a building in a single value. A hopefully exhaustive review of the existing indices for long-term evaluation of thermal comfort conditions in buildings and for thermal risk assessment is presented here for several reasons: (i) there are considerable differences in their structure and significance, (ii) some of them are based on thermal comfort models whilst others derive from rules of thumb and (iii) a systematic collection of those is missing. In this chapter, indices are first grouped according to the assumptions on which they are based on, and then they are analyzed in order to identify their scope, strengths and weaknesses. Finally, some guidance is provided for future developments.

1.1 Background

The relationship between physical parameters characterizing an indoor environment and the resulting human thermal perception has been studied by a number of authors. Many indices have been proposed over time, and terms like *discomfort index*, *stress index* or *heat index* have been commonly used in the scientific literature to identify metrics whose definition involves physiological, psychological, medical, climatological and engineering aspects.

Most of the studies aim to derive a single metric that, depending on the purpose, is able: (i) to identify possible thermal risk to human health and "discomfort-related events such as human morbidity and mortality, employee absenteeism, and crime" [1], (ii) to assess human thermal perception in a thermal environment—outdoor or indoor—and (iii) to assess climate. In general, a heat stress index is:

S. Carlucci, *Thermal Comfort Assessment of Buildings*,
PoliMI SpringerBriefs, DOI: 10.1007/978-88-470-5238-3_1,
© The Author(s) 2013

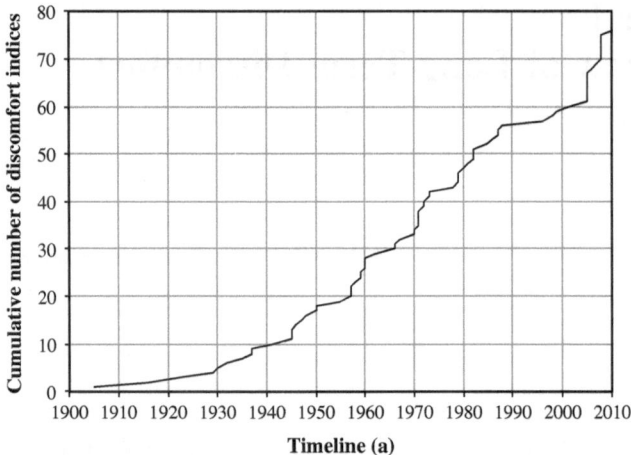

Fig. 1.1 The cumulative number of stress indices proposed over time

[…] a single value that integrates the effect of the basic parameters in any human thermal environment such that its value will vary with the thermal strain experienced by the individual [2].

In Fig. 1.1, the cumulated number of discomfort indices has been represented. Starting from a list by Epstein and Moran [3], more than seventy indices have been collected and reported in Appendix A. Also a number of reviews about them have been produced over time [3–5], and a detailed comparison of some of these indices can be found in [6].

Different classifications have been proposed to collect similar indices in homogeneous families. For example, most of the metrics were developed to evaluate summer overheating and just few of them deal with the assessment of winter discomfort [7, 8].

MacPherson [5] provided a useful classification, which groups the indices according to what they are based on: (i) calculation of the heat balance of the human body, (ii) physiological strain and (iii) measurement of physical parameters. More recently, a new type of indices has been developed in order to assess the long-term thermal comfort conditions in buildings. The present work focuses on these last ones.

1.1.1 Indices Based on the Heat Balance of the Human Body

These indices generally combine physiological parameters (e.g., the value of skin wetness and mean skin temperature), behavioral parameters (e.g., metabolic rate connected to activity and thermal resistance of clothing) and parameters of the thermal environment (e.g., dry-bulb air temperature, mean radiant temperature evaluated in a representative position of a room, relative humidity or wet-bulb temperature, air velocity, etc.). Most of these indices have been developed by relating the thermal sensations of people placed in thermostatic chambers wearing different clothing ensembles and performing different activities with a variety of environmental conditions (air temperature, humidity, air speed and mean radiant temperature). Fanger [9] proposed a famous comfort model based on two indices called *Predicted mean vote* (PMV) and *Predicted percentage of dissatisfied* (PPD).

1.1.2 Indices Based on Physiological Strain

These metrics were generally developed by correlating a wide range of environmental conditions and behavioral parameters with the thermal strain produced on individuals. The principle is "conditions of equal environmental stress are those which produce an equal physiological strain" [5]. This connection is usually expressed mathematically through multiple regression equations. Indices belonging to this family can depend on a number of different environmental parameters. The simplest ones, or their simplified versions, evaluate just up to two environmental parameters at a time and can be represented on psychometric charts through regions or zones (e.g, Effective temperature). Other examples of physiological strain indices[1] are the Effective temperature, the Corrected effective temperature, the Equatorial comfort index, the Index of physiological effect, the Predicted four-hour sweat rate and the Thermal strain index.

1.1.3 Indices Based on the Measurement of Physical Parameters

This family of indices is based on direct measurements of the physical parameters characterizing a thermal environment, generally analyzed through a linear regression model. The physiological effects are usually not taken into account. Indices of this family differ by the number and type of the environmental parameters considered and possibly by the values used for weighting the equivalent heat effect of each parameter. Examples of direct indices[2] are the Equivalent temperature, the Globe-thermometer temperature and the Operative temperature.

---

[1]   See Appendix A for references.

[2]   See Appendix A for references.

1.2 Indices for the Long-Term Evaluation of General Thermal Discomfort

These indices use simulated or measured data to evaluate long-term general thermal comfort conditions in buildings. A first overheating criterion based on the dry-resultant temperature was introduced by Chartered Institution of Building Services Engineers (CIBSE) [10]. Then, the International standard ISO 7730 [11] introduced five methods developed upon the Fanger comfort model and, more recently, three of such five indices were re-proposed in the European standard EN 15251 [12] extending their scope also to the adaptive comfort model, if compatible [13]. Later, Nicol et al. [14] introduced the Overheating risk, after having analyzed data measured in free-running buildings during the SCATs Project [15]. Another way to estimate heating or cooling loads or the indoor thermal comfort conditions during summer and winter consists in using degree-hours, which could be referred to different base temperatures. In some cases, these indices have been weighted in order to consider specified issues; for example, the index called Exceedance$_M$ weights discomfort hours by hourly average occupancy [16].

It is useful to collect into families the indices, which present homogeneous features. A proposal for their classification is presented in Table 1.1. Some of such indices are based on thermal comfort models codified in standards (the Fanger model primarily in ISO 7730, the EN adaptive model in EN 15251 and the ASHRAE adaptive model [17] in ASHRAE 55 [18]) and some of them are

Table 1.1 Families of the indices for the long-term evaluation of the general thermal comfort conditions in buildings

Family of indices	Based on comfort models				Based on reference temperatures		
	Year	Index name	Comfort model	Author(s)	Year	Index name	Author(s)
Percentage indices	2005	Percentage outside the (PMV) range	Fanger	ISO 7730	2002	CIBSE guide J criterion	CIBSE
	2005	Percentage outside the ($\ddot{o}_{i\text{op}}$) range	Fanger	ISO 7730	2006	CIBSE guide A criterion	CIBSE
	2005	Percentage outside the ($\ddot{o}_{i\text{op}}$) range	Adaptive EU	EN 15251			
	2005	PPD weighted criterion	Fanger	ISO 7730			
	2005	Accumulated PPD	Fanger	ISO 7730			
Cumulative indices	2005	Degree-hour criterion	Fanger	ISO 7730			
	2005	Degree-hour criterion	Adaptive EU	EN 15251			
	2010	Exceedance$_{PPD}$	Fanger	Borgeson-Brager			
	2010	Exceedance$_{Adaptive}$	Adaptive USA	Borgeson-Brager			
Risk indices	2008	Overheating risk	Adaptive EU	Nicol et al.	2007	Overheating risk	Robinson-Haldi
Averaging indices	2005	Average PPD	Fanger	ISO 7730			

constructed on the concept of comfort category (Table 1.2) or acceptability range (Table 1.3). Thermal comfort categories are defined in different ways in the standards. ISO 7730 proposed three categories called A, B and C without specifying their scope. Their boundaries are expressed in terms of PMV and are the same in value of those of the EN 15251: for Category A, PMV is comprised in the interval [−0.2, +0.2]; for Category B, in the interval [−0.5, +0.5] and for Category C, in the interval [−0.7, +0.7].

The target comfort category or acceptability range has to be chosen on the base of (i) the level of thermal comfort acceptability required in a building for ASHRAE 55, (ii) the typology of intervention—construction of a new building or refurbishment—for EN 15251 and (iii) if very low variations of indoor environmental parameters are required for EN 15251 and ASHRAE 55.

Table 1.2 Definition of the comfort categories according to EN 15251

EN 15251 category	Description	Fanger		Adaptive
		PPD (%)	PMV	$\Delta\theta_{op}$ (K)
I	High level of expectation and is recommended for spaces occupied by very sensitive and fragile persons with special requirements like handicapped, sick, very young children and elderly persons	≤ 6	$-0.2 \leq PMV \leq +0.2$	± 2
II	Normal level of expectation and should be used for new buildings and renovations	≤ 10	$-0.5 \leq PMV \leq +0.5$	± 3
III	An acceptable, moderate level of expectation and may be used for existing buildings	≤ 15	$-0.7 \leq PMV \leq +0.7$	± 4
IV	Values outside the criteria for the above categories. This category should only be accepted for a limited part of the year	>15	$PMV < -0.7$ or $PMV > 0.7$	

Table 1.3 Definition of the comfort categories according to ASHRAE 55

ASHRAE 55 class (%)	Scope	PPD (%)	Fanger PMV	Adaptive $\Delta\theta_{op}$ (K)
90	It shall be used when a higher standard of thermal comfort is desired	≤ 10	$-0.5 \leq PMV \leq +0.5$	± 2.5
80	It is for typical applications and shall be used when other information is not available	≤ 20	$-0.85 \leq PMV \leq +0.85$	± 3.5

The selection of a thermal comfort category implies that actual or simulated indoor operative temperatures have to be included in an interval of (i) operative temperatures, constituted by the offset for a certain number of degrees (°C or °F) above and below the theoretical comfort temperature for the adaptive comfort models and (ii) PMV, based on actual or simulated/assumed indoor and personal parameters (clothing levels and metabolic activity) constituted by the offset for a certain value of PMV above and below the neutrality (PMV $= 0$) for the Fanger comfort model. For more detailed information see [19].

Long-term discomfort indices are described according to a number of key aspects (Table 1.4) in order to highlight similarities and differences, and a description of every key aspect follows.

1.2.1 Indices Based on Comfort Models

Some long-term discomfort indices are based on comfort models (Fanger and adaptive) referring to the comfort operative temperature (e.g., Nicol et al.'s Overheating Risk) or to a specified comfort category (e.g., Percentage Outside Range). By contrast, other indices are not based on any explicit comfort model; for example, they compare indoor operative temperature with a fixed reference temperature (e.g., CIBSE criteria). Such fixed reference temperatures can be calculated from hourly weather data or are just stated by the method; but, in any case, they do not refer to a comfort model.

1.2.2 Category-Dependent Indices

The indices based on comfort models differ whether they rely on comfort categories. According to category-dependent indices, thermal discomfort is caused by the exceedance outside the boundaries of a given comfort category. They generally have weighting factors that are zero inside the category and non-zero outside. Therefore, category-dependent indices will assume different values if the category changes, even if the climate, the building, its use and time series either of indoor operative temperature or of the PMV are the same. This fact could result in a source of misunderstanding if the comfort category is not made explicit together with the value of the index.

1.2.3 Symmetric and Asymmetric Indices

The actual indoor operative temperature can result to be above or below a supposed comfort temperature in a specified period of the year. This phenomenon is commonly considered uncomfortable and it is called overheating when the indoor operative temperature exceeds the comfort temperature and overcooling when it is lower. Some indices evaluate discomfort just accounting for overheating during summer and overcooling during winter; while others take into account both. We call the former ones asymmetric indices, the latter symmetric indices. The difference between them becomes visible with the following example: during summertime, overheating is typically prevalent, but in some hours during the night, operative temperature can be below a given comfort temperature. This results in an uncomfortable condition only according to symmetric indices.

1.2.4 Indices Applicable Just to Summer or Extensible also to Winter

Most long-term discomfort indices are exclusively proposed for predicting discomfort caused by summer overheating. Some of them are based on assumptions which exclusively refer to the summer period, so they must not be used during winter or, in general, cold periods. Other indices instead can be also used for the comfort assessment of cold periods since they are obtained by comparing the actual or simulated indoor conditions with a reference operative temperature or PMV during a specified period, which may be warm or cold (i.e., summer or winter).

1.2.5 Discomfort Scales and Thresholds

Some of the long-term discomfort indices are accompanied by arbitrary rules of thumb on what levels of discomfort might be acceptable in a building. These rules are usually proposed as thresholds. Such thresholds are expressed in the same scale of units of the indices: some range from zero to plus infinity, other are percentages. It should be noted that the former do not have an upper limit and this could be a limit when comparing the comfort performances of buildings under different climatic or operational conditions; the latter derive from a normalization, where the total number of occupied hours in a given period is usually the term for the normalization.

1.3 Description of the Long-Term Discomfort Indices

In case of a multi-zone building, EN 15251 is not univocal: in the Sect. 8.3 *Calculated indicators of indoor environment* states:

> [...] the building meets the criteria of a specific category if the rooms representing 95 % of building volume meet the criteria of the selected category [12].

whilst in the Annex I at the Sect. 1.2 *Whole year computer simulations of the indoor environment and energy performance*, it states:

Table 1.4 Summary of the features for every long-term discomfort index

Family of indices	Long-term discomfort indices	Index based on a comfort model	Index dependent on comfort categories	Symmetric index	Index applicable only to summer	Index provided with a comfort threshold
Percentage indices	Percentage outside the (PMV) range	F	√	√		√
	Percentage outside the $(\theta_{op,\ PMV})$ range	F	√	√		√
	Percentage outside the $(\theta_{op,\ Adaptive})$ range	A_{EU}	√	√		√
	CIBSE guide J criterion				√	√
	CIBSE guide A criterion				√	√
	PPD weighted criterion	F	√			
	Accumulated PPD	F		√		
Cumulative indices	Degree-hour criterion	F	√			
	Degree-hour criterion	A_{EU}	√			
	Exceedance$_{PPD}$	F	√		√	
	Exceedance$_{Adaptive}$	A_{USA}	√		√	
Risk indices	Nicol et al.'s Overheating risk	A_{EU}			√	
	Robinson-Haldi's Overheating risk			√	√	√
Averaging indices	Average PPD	F		√		

F Fanger model, A_{EU} EN adaptive model, A_{USA} ASHRAE adaptive model

It is then calculated how the temperatures are distributed between the 4 categories. This is done by a floor area weighted average for 95 % of the building spaces [12].

Therefore, EN 15251 indicates to weight for the building volume, but it uses the floor area in the Annex I (it is also not explicitly specified if it respectively refers to the net conditioned volume and net floor area). Of course, if the height inside each zone of the building is constant, the two correspond. Instead, ISO 7730 does not specify how to calculate long-term discomfort indices in case of a multi-zone building.

At a given moment, the exceedance of the comfort limits in a multi-zone building is a distribution rather than a single value. In other words, even in correspondence to the same outdoor climate conditions, the building performs differently in different spaces. Kalz and Pfafferott [20], assuming that such exceedance is normally distributed over all zones of a building, recommend, during monitoring campaigns, to use the floor area weighted average for 84 % of the building spaces where the hourly indoor temperatures remains within the defined comfort limits.

1.3.1 Percentage Indices

Comfort performance of a building is assessed, for this family of indices, by calculating the ratio of likely discomfort hours to the total number of occupied hours. They differ in the proxy used to describe comfort conditions: some use a comfort model, others a reference temperature.

1.3.1.1 Percentage Outside the Range

The method called Percentage outside the range (POR), introduced by ISO 7730 and re-proposed by EN 15251, requires to calculate the percentage of hours of occupation when the—actual or simulated—PMV or indoor operative temperature are outside a specified comfort range related to the chosen comfort category (see Table 1.2)

$$POR \equiv \frac{\sum_{i=1}^{Oh}(wf_i \cdot h_i)}{\sum_{i=1}^{Oh} h_i} \in [0,1]. \tag{1.1}$$

It is suitable for both Fanger's and adaptive comfort models. When referring to the Fanger model and the comfort range is expressed in terms of PMV, the index is here indicated with the term $POR_{\text{Fanger,PMV}}$

$$POR_{PMV}^{Fanger} \propto \begin{pmatrix} wf_i = 1 \Leftarrow (PMV < PMV_{\text{lower limit}}) \vee (PMV > PMV_{\text{upper limit}}) \\ wf_i = 0 \Leftarrow (PMV_{\text{lower limit}} \leq PMV \leq PMV_{\text{upper limit}}) \end{pmatrix}. \tag{1.2}$$

By making assumptions about variables affecting PMV (clothing, metabolic rate, air velocity, relative humidity and the relationship between indoor dry-bulb air and mean radiant temperatures) and introducing a relevant level of uncertainty, the PMV boundaries of a selected category can be translated into operative temperature: ISO 7730 reports it in Appendix E and EN 15251 in Table A.3. Thus, the actual operative temperature can be compared to the category limits, and the index is indicated with $POR_{\text{Fanger},\theta\text{op}}$ and

$$POR_{\theta_{op}}^{Fanger} \propto \begin{pmatrix} wf_i = 1 \Leftarrow (\theta_{\text{op, actual PMV}} < \theta_{\text{lower limit}}) \vee (\theta_{\text{op, actual PMV}} > \theta_{\text{upper limit}}) \\ wf_i = 0 \Leftarrow (\theta_{\text{op, lower limit}} \leq \theta_{\text{op, actual PMV}} \leq \theta_{\text{op, upper limit}}) \end{pmatrix}. \tag{1.3}$$

When referring to the adaptive model, the comfort range is expressed in terms of operative temperatures. In this case, the index is indicated with POR_{Adaptive} and its formulation is

$$POR^{Adaptive} \propto \begin{pmatrix} wf_i = 1 \Leftarrow \left(\theta_{op,\,in} < \theta_{op,\,lower\,limit}\right) \vee \left(\theta_{op,\,in} > \theta_{op,\,upper\,limit}\right) \\ wf_i = 0 \Leftarrow \left(\theta_{op,\,lower\,limit} \leq \theta_{op,\,in} \leq \theta_{op,\,upper\,limit}\right) \end{pmatrix}.$$

$$(1.4)$$

The POR appears to be a straightforward and simple method, to compare the comfort performance of different buildings in different climates According to its definition, this index is comfort model-based, category-dependent and symmetric (since it considers both the upper and lower exceedance outside the boundaries). Moreover, it can be theoretically applied to summer and winter, and it is also completed by a proposal for a maximum discomfort threshold, as EN 15251 states:

> The parameter in the rooms representing 95 % of the occupied space is not more than as example 3 % (or 5 %) of occupied hours a day, a week, a month and a year outside the limits of the specified category [12].

On the other hand, POR does not give information about the severity of the uncomfortable conditions and introduces a discontinuity at the boundaries of the category that has no correspondence with physics and physiology.

1.3.1.2 Percentage of Occupied Hours Above a Reference Temperature

Chartered Institution of Building Services Engineers (CIBSE) proposed two design-overheating criteria to be used when performing building simulation using the *Design Summer Year* (DSY)

$$CIBSE_{A,\,J} \equiv \frac{\sum_{i=1}^{Oh} \left(wf_i^{A,J} \cdot h_i\right)}{\sum_{i=1}^{Oh} h_i} \in [0,1].$$

$$(1.5)$$

They use the Dry-resultant temperature defined as the temperature recorded by a thermometer at the center of a blackened globe of 100 mm diameter [21]

$$\theta_{res} = \frac{\theta_{mr,\,in} + \theta_{db,\,in}\sqrt{10 \cdot v_{a,in}}}{1 + \sqrt{10 \cdot v_{a,in}}} \approx \left. \frac{\theta_{mr,\,in} + \theta_{db,\,in}}{2} \right|_{v_{a,\,in}\,\leq\,0.1\,ms^{-1}}$$

$$(1.6)$$

The first of the two metrics, called here CIBSE$_J$, requests that dry-resultant temperature should not exceed 25 °C for more than 5 % of the occupied time [10]

$$\begin{cases} CIBSE_j \propto \begin{pmatrix} wf_i^J = 1 \Leftarrow \left(\theta_{res} > 25°C\right) \\ wf_i^J = 0 \Leftarrow \left(\theta_{res} \leq 25°C\right) \end{pmatrix}. \\ CIBSE_J \leq 0.05 \end{cases}$$

$$(1.7)$$

The second one, called here CIBSE$_A$, tolerates the exceedance of a dry-resultant temperature equal to 28 °C for not more than 1 % of occupied hours in naturally ventilated buildings, except for bedrooms where a lower threshold of 26 °C is specified [22]

$$\begin{cases} CIBSE_A \propto \begin{pmatrix} wf_i^A = 1 \Leftarrow (\theta_{res}^{\text{bedrooms}} > 26°C) \vee (\theta_{res}^{\text{other rooms}} > 28°C) \\ wf_i^A = 0 \Leftarrow (\theta_{res}^{\text{bedrooms}} \leq 26°C) \vee (\theta_{res}^{\text{other rooms}} \leq 28°C) \end{pmatrix}. \\ CIBSE_A \leq 0.01 \end{cases} \quad (1.8)$$

They are asymmetric indices; in fact they are only used for assessing the overheating likelihood. They are not related to comfort models and do not depend on comfort categories.

Nicol et al. [14] identified some problems in using a fixed threshold temperature and a simple exceedance criterion in free-running buildings: (i) data and analysis show that, in these buildings, the comfort temperature is not a fixed value but varies with the outdoor dry-bulb temperature (in the form of running-mean or monthly mean); (ii) indices based only on the hours of exceedance are not able to provide the magnitude, but just the occurrence, of overheating, and (iii) methods based on a threshold temperature are sensitive to the assessment method of the indoor temperatures.

1.3.2 Cumulative Indices

According to this family of indices, accumulation of thermal stress during the occupied period results in perceived discomfort. They are not expressed as percentages. They differ from each other in using or not a comfort model and how they assess the weight of the hourly thermal stress.

1.3.2.1 PPD-Weighted Criterion

Only proposed for the Fanger comfort model, the PPD-weighted criterion (PPDwC) was introduced for the first time in 2001 by Olesen et al. [23]. It consists of weighting the time during which PMV exceeds the comfort boundaries with a weighting factor, wf_i, and then calculating the index by the summation hour-by-hour of the products of wf_i per time

$$PPDwC \equiv \sum_{i=1}^{Oh} (wf_i \cdot h_i) \in [0, +\infty). \quad (1.9)$$

Warm and cold periods are evaluated separately, thus this index is asymmetric

$$\begin{cases} PPDwC_{\text{warm period}} \Leftrightarrow (PMV > PMV_{\text{upper limit}}) \\ PPDwC_{\text{cold period}} \Leftrightarrow (PMV < PMV_{\text{lower limit}}) \end{cases} \quad (1.10)$$

where $PMV_{\text{lower limit}}$ and $PMV_{\text{upper limit}}$ are respectively the lower and upper limits of a specified comfort range (see Table 1.2). The calculation of the wf_i is different in EN 15251 and ISO 7730.

According to EN 15251:

$$PPDwC^{\text{EN 15251}} \propto \left(\begin{array}{l} wf_i \equiv \dfrac{PPD_{\text{actual PMV, i}}}{PPD_{\text{PMV limit}}} \Leftarrow \left(PMV > PMV_{\text{upper limit}} \right) \\[3mm] wf_i = 0 \Leftarrow \left(PMV_{\text{lower limit}} \leq PMV \leq PMV_{\text{upper limit}} \right) \\[3mm] wf_i \equiv \dfrac{PPD_{\text{actual PMV, i}}}{PPD_{\text{PMV limit}}} \Leftarrow \left(PMV < PMV_{\text{lower limit}} \right) \end{array} \right) \quad (1.11)$$

while, according to ISO 7730

$$PPDwC^{\text{ISO 7730}} \propto \left(\begin{array}{l} wf_i \equiv \dfrac{PPD_{\text{actual PMV, }i}}{PPD_{\text{PMV limit}}} \Leftarrow \left(|PMV| > |PMV_{\text{limit}}| \right) \\[3mm] wf_i = 1 \Leftarrow \left(PMV = PMV_{\text{limit}} \right) \end{array} \right). \quad (1.12)$$

Excluding the different formulation of the domain of the function wf_i, the sole difference between them is that ISO 7730 increases the value of the index when the actual PMV is equal to the PMV limits. This distinction would be negligible if compared with the effect due to considering uncertainty with which PMV is estimated, based on the uncertainty which affects the individual input parameters. It is, in fact, a good opportunity to underline here that PMV estimate is indeed affected by uncertainty as any other quantity measured or derived from measurements [24].

An interesting feature of this index (in both the versions) is that long-term thermal discomfort is an explicit function of the estimated percentage of dissatisfied. The limits of this method, in the current formulations, are: (i) it cannot be applied for assessing free-running buildings, (ii) it does not consider the effect of overcooling during the summer and overheating during winter, (iii) the index cannot be plotted on a fixed scale and (iv) it heavily depends on the boundaries of comfort categories, and those are subject to a certain debate in the scientific literature [24, 25].

1.3.2.2 Accumulated PPD

Accumulated PPD (Sum_PPD) was introduced by ISO 7730 and exclusively proposed for the Fanger model. It is defined as the summation, with a time step of one hour, of all PPDs over the occupied hours, thus not depending on comfort categories and being a symmetric index

$$Sum\_PPD \equiv \sum_{i=1}^{Oh} PPD_i \in [0, +\infty). \quad (1.13)$$

Since it is a simple summation of percentages over time it has the unit of measure of a percentage, but it ranges from zero to *plus-infinitum*. Thus, even if it uses the predicted percentage of dissatisfied, (i) it is not able to allow a comparison of the severity of possible thermal conditions inside a building: the same value can be reached summing few hours with high PPD or adding a larger number of hours with lower predicted uncomfortable conditions. Other limits are: (ii) it can only be applied with the Fanger comfort model, (iii) it cannot be plotted on a fixed scale, e.g., the comfort footprint, and (iv) it is difficult to define rules for assigning comfort performances.

1.3.2.3 Degree-Hours Criterion

Degree-hours criterion (DhC) was introduced by ISO 7730 and was subsequently included with some modifications in EN 15251. According to it, the hours when actual operative temperature exceeds the specified comfort range during occupied hours are weighted by a factor called *wf*. This weighting factor depends on the module of the difference, at a certain hour, between actual or calculated operative temperature, θ_{op}, and the lower or upper limit, $\theta_{op,\ limit}$, of a specified comfort range. Necessarily, if comfort range is specified in terms of PMV, the limits have to be translated into operative temperature by making assumptions on clothing, metabolic activity, air velocity, relative humidity and the relationship between air and mean radiant temperatures.

The weighting factor, *wf*, has two different formulations according to the two aforementioned standards

$$\begin{cases} wf_i\Big|_{Fanger}^{\text{ISO 7730}} \equiv 1 + \dfrac{\left|\theta_{op,\ i}-\theta_{op,\ \text{limit}}\right|}{\left|\theta_{op,\ \text{comfort}}-\theta_{op,\ \text{limit}}\right|} \\ wf_i\Big|_{Fanger}^{\text{EN 15251}} \equiv \left|\theta_{op,\ i} - \theta_{op,\ \text{limit}}\right| \end{cases}. \tag{1.14}$$

The ISO formulation (i) penalizes the recurrence of the exceedance since, assuming two situations characterized by the same accumulation number of degrees outside the comfort range but distributed differently over time, the situation with more discomfort hours has a higher value of the index and (ii) weights the discomfort by the amplitude of comfort range depending on the selected comfort category: the value of the index is indirectly proportional with respect to the semi-amplitude of the comfort range (denominator), thus, for the same exceedance in degrees (numerator), the calculated discomfort index is higher in those situations where stricter thermal conditions are requested, for example, in building falling into Category A. On the contrary, the EN formulation defines the weighting factor as the summation of the hourly exceedance of operative temperature outside the boundaries of a specified comfort range. Such formulation can be easily extended also to the adaptive comfort model.

$$wf_i\Big|_{Adaptive}^{\text{EN 15251}} \equiv \left|\theta_{op,\ i} - \theta_{op,\text{limit}}\right|. \tag{1.15}$$

For a characteristic period (usually a year or a season) the index is calculated by summing the products of the weighting factor and time (usually expressed in time steps of one hour).

$$DhC \equiv \sum_{i=1}^{Oh} (wf_i \cdot h_i) \in [0, +\infty). \tag{1.16}$$

For both the standards in the warm period (summer, according to EN 15251), the summation is extended exclusively to occupied hours when $\theta_{op,i} > \theta_{op, \text{upper limit}}$. Similarly for the cold period (winter, according to EN 15251), the summation is extended only to occupied hours when $\theta_{op,i} < \theta_{op, \text{lower limit}}$, thus both the standards propose an asymmetric use of DhC.

$$\begin{cases} DhC_{\text{Warm period}} \propto \begin{pmatrix} wf_i > 0 \Leftarrow (\theta_{op,\,i} > \theta_{op, \text{upper limit}}) \\ wf_i = 0 \Leftarrow (\theta_{op,\,i} \leq \theta_{op, \text{upper limit}}) \end{pmatrix} \\ DhC_{\text{Cold period}} \propto \begin{pmatrix} wf_i > 0 \Leftarrow (\theta_{op,\,i} < \theta_{op, \text{lower limit}}) \\ wf_i = 0 \Leftarrow (\theta_{op,\,i} \geq \theta_{op, \text{lower limit}}) \end{pmatrix} \end{cases} \tag{1.17}$$

This index is based on the comfort models and is also category-dependent. A limit of this index is that it cannot be plotted on an percentage scale and it is not normalized to the total number of occupied hours: the comparison of buildings characterized by different number of occupation hours can be misunderstood. As all indices based on categories, it introduces a discontinuity at the boundaries of the category, which has no correspondence with physics and physiology.

1.3.2.4 Exceedance$_M$

Exceedance$_M$ was proposed by Borgeson and Brager [16]. It predicts thermal discomfort when the indoor operative temperature is higher than the reference given by the upper limit of the 80 % acceptability range during the summer occupied hours (Oh). Each discomfort hour is weighted with the number of people in a given zone in the same hour. Then, their summation over the calculation period is divided by the total number of occupants present in the zone during all discomfort hours of the calculation period

$$Exceedance_M \equiv \frac{\sum_{i=1}^{Oh} (n_i \cdot B_i)}{\sum_{i=1}^{Oh} n_i} \in [0, 1] \tag{1.18}$$

$$Exceedance_M \propto \begin{pmatrix} B_i = 1 \Leftarrow (\text{Acceptability} < 80\,\%) \\ B_i = 0 \Leftarrow (\text{Acceptability} \geq 80\,\%) \end{pmatrix} \tag{1.19}$$

where n_i is the number of occupants inside a given thermal zone at a certain hour i, B_i is a binary variable that indicates whether comfort conditions are exceeded.

Exceedance$_M$ is an asymmetric index proposed only for the summer assessment against overheating. It depends on comfort categories and can be calculated using both the Fanger and the ASHRAE adaptive comfort models (the index is called respectively Exceedance$_{PPD}$ and Exceedance$_{Adaptive}$).

The Exceedance$_M$ is the only index among the ones reviewed that is calculated using the occupation rate. This seems a fascinating feature that aims at overcoming the hard cutoff due to the occupancy binary code 0 for an unoccupied hour and 1 for an occupied hour and might be useful to be incorporated in other indices. On the other side, the hard cutoff at 80 % acceptability can cause an anomalous evaluation if consecutive values are just little up or down of the specified threshold. Moreover, this index does not take into account the severity of the thermal exceedance; the weight B_i is equal to 1 irrespectively of how large is the difference between actual operative temperature and the upper comfort threshold.

1.3.3 Risk Indices

Risk indices aim at assessing the likelihood that a discomfort phenomenon could happen, given specified conditions of the indoor environment.

1.3.3.1 Nicol et al.'s Overheating Risk

Introduced in [14], Nicol et al.'s overheating risk (NaOR) assumes that thermal discomfort is related to the difference ($\Delta\theta$) between the actual operative temperature and the EN adaptive comfort temperature, and not just to a fixed threshold.

Their analysis of comfort questionnaires collected in European naturally ventilated office buildings during the SCATs Project [15] confirms that some people may feel uncomfortable even at the theoretical comfort (neutral) temperature. When the theoretical comfort temperature is exceeded, the offset is recorded, and a weighting factor is calculated on an hourly basis. The weighting factor expresses the non-linearity between thermal discomfort and the exceedance from the theoretical comfort temperature [26].

The authors derived the likelihood of overheating, $P(\Delta\theta)$, from a logistic regression analysis. The index predicts the percentage of individuals voting +2 or +3 (i.e., warm or hot) on the ASHRAE thermal sensation scale:

$$P(\Delta\theta) \equiv \frac{\exp(0.4734\Delta\theta - 2.607)}{1 + \exp(0.4734\Delta\theta - 2.607)} \in [0.07, 1.00]. \qquad (1.20)$$

NaOR was derived from the same comfort database as EN 15251 adaptive comfort model and it is not related to comfort categories. It is an asymmetric index which aims at predicting overheating phenomena and cannot be applied in mechanically cooled buildings.

1.3.3.2 Robinson's and Haldi's Overheating Risk

Robinson and Haldi [27] base their overheating risk index, called here Robinson's and Haldi's overheating risk (*RHOR*), on the assumption that the charge and discharge of human tolerance to overheating may be modeled with the behavior of an electrical capacitor. Thermal satisfaction is thus related to accumulated overheating stimuli instead of instantaneous satisfaction. In other words, the authors hypothesize that occupants are tolerant of occasional offsets from comfort conditions and overheating is due to an accumulation of heat stress events rather than a single event. According to this analytical model, the heat stimulus is related to the exceedance of a reference temperature of 25 °C, and it can be calculated with the equations

$$
\begin{cases}
P_{OH}^{Charging}(t_n) \equiv 1 - \exp\left(-\alpha \sum_{i=1}^{n} Dh_{t_0,t_i}\right)(1 - P_{OH}(t_0)) \in [0,1] \\
P_{OH}^{Discharging}(t_n) \equiv \exp\left(-\beta \sum_{i=1}^{n} Dh_{t_0,t_i}^*\right) P_{OH}(t_0) \in [0,1]
\end{cases}
\tag{1.21}
$$

where $P_{OH}(t_n)$ is the probability of overheating at time n, $P_{OH}(t_0)$ is the probability of overheating at the time of transition from charging to discharging, Dh are the number of the degree-hours above the reference temperature during the calculation period, and Dh^* are the number of the degree-hours below the reference temperature (25 °C) during the calculation period. The two coefficients 'α'—a time constant for charging—and 'β'—a time constant for discharging– have to be tuned empirically in relation to the specific situation. The authors suggest a threshold for the probability of overheating of 20 % since: "current thermal comfort standards target a PPD of $\leq$20 %" [27].

However this model: (i) is thought for a continuous period and not just for occupied hours, (ii) in order to evaluate comfort conditions, considers just the indoor dry-bulb air temperature, while it neglects clothing and activity adaptation or occupants' ability to control the indoor environment (iii) has also been calibrated only for a temperate climate, (iv) is based on a reference temperature of 25 °C, which was chosen without a justification based on existing comfort models or independent observations and (v) the time constants for charging and discharging need a big number of observations to be tuned (the value for the constant 'β' was just assumed and not derived from measurements).

1.3.4 Averaging Indices

This family of long-term discomfort indices, consisting of just one metric, estimates the likelihood of thermal discomfort simply averaging a short-term discomfort index over an occupation period.

1.3.4.1 Average PPD

Introduced by ISO 7730 and not included in EN 15251, Average PPD ($<PPD>$) consists in calculating the mean PPD over the occupied hours

$$<PPD> \equiv \frac{\sum_{i=1}^{Oh} (PPD_i \cdot h_i)}{\sum_{i=1}^{Oh} h_i}. \qquad (1.22)$$

It is an index based on a comfort model, but it does not depend on comfort categories. In the way it is defined, it is a symmetric index.

Its most notable feature is that, unlike indices based on categories, it does not introduce discontinuities since the actual discomfort stress is assessed through the *Predicted percentage of dissatisfied*, which is a continuous and smooth function. It can be used for comfort optimization procedures and for comparing the thermal comfort performance of different buildings. However, since it relies only on the Fanger model, hence it would not respond to the needs of designers/researchers willing to design/evaluate a naturally ventilated building by using an adaptive comfort model.

1.4 Conclusions

None of the reviewed indices showed to be fully suitable for the long-term evaluation of the general thermal comfort conditions in a building. All boundary conditions that affect their calculation should be made explicit to produce reliable results, which can be clearly interpreted. Also, when comparing the thermal comfort performance of different buildings or building variants, the boundary conditions must be harmonized among buildings/variants (e.g., through the coding in a standard) while, for example, ISO 7730 and EN 15251 do not propose any methodology to define the duration of warm or cold periods. This is a serious limit of both the standards since the long-term discomfort indices vary considerably in consequence of variations of the calculation period. Nicol et al. wrote:

> Merely increasing the hours of occupation may 'solve' an overheating problem, (...), which is clearly unrealistic [14].

Moreover, these standards do not include a harmonized common method for calculating the overall index for a multi-zone building. EN 15251 suggests calculating the indices in 95 % (or 97 %) of the space and weighting the indices calculated for each zone for the respective air volume. This approach could have the advantage of being simple, but, since each zone can have a different occupation rate, it does not estimate the probability that the occupants can undergo uncomfortable conditions. An option to overcome this problem and also the hard cutoff due to the occupancy binary code (0 for an unoccupied hour and 1 for an occupied hour) could be to weight discomfort index of each zone for the respective occupation rate (following the example of the Exceedance$_M$).

Furthermore, clearer and more precise rules for zoning of large or complex buildings are essential if the long-term discomfort indices have to be used for assessing thermal comfort conditions in existing buildings or as guidance for design. On this issue also, agreed guidelines for zoning have to be proposed and hopefully included in a standard.

Comfort categories, on which some of the reviewed indices are based, can be affected by some limitations: (i) they differ depending on what standard (ASHRAE or EN) is considered; (ii) Arens et al. [25] argue that the range proposed for the Category I of EN 15251 (and for the Category A of ISO 7730) is too narrow to be detected by the occupants of a building; (iii) using Alfano et al.'s words: "the PMV range required by A-category can be practically equal to the error due to the measurements accuracy and/or the estimation of parameters affecting the index itself" [24]. The same ISO 7730 admits an objective difficulty in measuring the input parameters for verification that PMV conforms to A-category requirement and suggests referring to equivalent operative temperature ranges. However, this means that the uncertainties on all other variables beside air and mean radiant temperatures (or globe-thermometer temperature) are not taken into account [19].

Uncertainties affecting the evaluation of the PMV, due to the errors either in measuring and estimating the physical and personal variables respectively, has also an implication on how sharp can be the change of the weighting factors on the boundaries of the categories. The precision by which PMV can be assessed has to be discussed for every calculation rule based on the comparison between the actual value and the boundary values of a certain comfort category expressed in terms of PMV (e.g., Eq. 1.11). In response to this and other problems, Nicol and Wilson [28] propose to relate the selection of the comfort category exclusively with the expectations of the occupants.

The indices based on the percentage of time outside the comfort range, (i) suffer for an abrupt change in the assessed comfort perception due to the step function that determines when a likely condition of discomfort occurs; (ii) they measure just the frequency of overheating, not its magnitude.

As for fixed reference temperature based indices, even if the reference temperature is associable to a theoretical comfort temperature, e.g., derivable from the Fanger comfort model according to specified assumptions, they may be reliably and exclusively used for assessing comfort in mechanically conditioned buildings and only if such assumptions are valid. For example, in naturally ventilated buildings, where summer theoretical comfort temperature varies with the outdoor dry-bulb temperature (monthly average or running-mean), a fixed overheating threshold might not be reliable for estimating the likelihood of discomfort.

1.5 Improvement Objectives

In summary, none of the considered indices for long-term discomfort altogether fulfills all desirable features. Each method has its pros and cons.

Percentage outside range can be applied to both the Fanger and the adaptive comfort models and evaluates both upper and lower exceedance from theoretical comfort temperature.

Averaged PPD, though not included in EN 15251, interestingly takes into account the actual thermal response of the occupants. On the other hand, it implies necessarily to use the Fanger comfort model, which might not be the choice of the designer or evaluator when dealing with a naturally ventilated building.

Nicol et al.'s overheating risk adapts well to assess free-running buildings in summer, but it is not applicable in mechanically cooled buildings. Also, it does not provide a unique value, but a time series of values with an hourly resolution.

Exceedance$_M$, unlike all other reviewed indices, takes into account the number of occupants in evaluating the likely thermal discomfort.

Therefore, having the objective to use a discomfort index for optimization purposes, it would be necessary to construct a new index:

- That is applicable for both free-running and mechanically cooled buildings and can be used with both the adaptive and the Fanger comfort models (e.g., *Percentage outside range*).
- That reflects the nonlinear relationship between perception of discomfort and the exceedance from the theoretical comfort temperature (e.g., *Averaged PPD* and *Nicol's overheating risk*).
- That, in case of multi-zone buildings, weights the zone indices by the number of occupants inside each zone, rather than by buildings features such as net volume, or net floor area (e.g., *Exceedance$_M$*).
- That is applicable to the evaluation of both summer and winter discomfort.
- That is symmetric, i.e., able to estimate possible discomfort due to the upper and lower exceedance from the theoretical comfort temperature.
- That is independent of discontinuities connected to the use of comfort categories.

References

1. L.S. Kalkstein, K.M. Valimont, An evaluation of summer discomfort in the United States using a relative climatological index. Bull. Am. Meteorol. Soc. **7**, 842–848 (1986)
2. K. Parsons, *Human Thermal Environments*, 2nd edn. (Taylor & Francis, London, 2003)
3. Y. Epstein, D. Moran, Thermal comfort and the heat stress indices. Ind. Health **4**, 388–398 (2006)
4. R. Quayle, F. Doehring, Heat stress, a comparison of indices. Weatherwise **34**, 120–124 (1981)
5. K. MacPherson, The assessment of the thermal environment a review. Br. J. Ind. Med. **19**, 151–164 (1962)

6. D.H.K. Lee, Seventy-five years of searching for a heat index. Environ. Res. **22**, 331–356 (1980)
7. M.Y. Beshir, J.D. Ramsey, Heat stress indices: a review paper. Int. J. Ind. Ergon. **3**, 89–102 (1998)
8. A. Court, Windchill over 40 years. Preprint volume of Extended Abstracts 1981: Fifth Conference on Biometeorology (American Meteorological Society, Anaheim, 1981), pp. 82–85
9. P.O. Fanger, *Thermal Comfort* (Danish Technical Press, Copenhagen, 1970)
10. CIBSE, *Guide J: Weather, Solar and Illuminance Data* (Chartered Institution of Building Services Engineers, London, 2002)
11. ISO:ISO 7730, *Ergonomics of the Thermal Environment: Analytical Determination and Interpretation of Thermal Comfort Using Calculation of the Pmv and Ppd Indices and Local Thermal Comfort Criteria*, 3rd edn. (International Standard Organization, Geneva, 2005)
12. CEN: EN 15251, *Indoor Environmental Input Parameters for Design and Assessment of Energy Performance of Buildings Addressing Indoor Air Quality, Thermal Environment, Lighting and Acoustics* (European Committee for Standardization, Brussels, 2007)
13. M.A. Humphreys, J.F. Nicol, Understanding the adaptive approach to thermal comfort. ASHRAE Trans. **104**(1), 991–1004 (1998)
14. J.F. Nicol, J. Hacker, B. Spires, H. Davies, Suggestion for new approach to overheating diagnostic, in *Proceedings of Conference: Air Conditioning and the Low Carbon Cooling Challenge* (Cumberland Lodge, Windsor, 2008)
15. J.F. Nicol, K. McCartney, *Final report of Smart Controls and Thermal Comfort (SCATs) Project Report to the European Commission of the Smart Controls and Thermal Comfort project.* (Oxford Brookes University, UK, 2001)
16. S. Borgeson, G.S. Brager, Comfort standards and variations in exceedance for mixed-mode buildings. Build. Res. Inf. **39**(2), 118–133 (2010)
17. R.J. de Dear, G.S. Brager, Developing an adaptive model of thermal comfort and preference. ASHRAE Trans. **104**(1A), 145–167 (1998)
18. ASHRAE: ANSI/ASHRAE 55, *Thermal Environmental Conditions for Human Occupancy.* (American Society of Heating, Refrigerating and Air-Conditioning Engineers, Atlanta, 2004)
19. L. Pagliano, P. Zangheri, Comfort models and cooling of buildings in the Mediterranean zone. Adv. Build. Energy Res. **4**, 167–200 (2010)
20. D.E. Kalz, J. Pfafferott, Comparative evaluation of natural ventilated and mechanical cooled non-residential buildings in Germany: thermal comfort in summer, in *Proceedings of Conference: Adapting to Change—New Thinking on Comfort* (Cumberland Lodge, Windsor, 2010)
21. CIBSE, *Guide A: Environmental Design.* (Chartered Institution of Building Services Engineers, London, 1999)
22. CIBSE, *Guide A: Environmental Design.* (Chartered Institution of Building Services Engineers, London, 2006)
23. B.W. Olesen, Introduction to the new revised draft of EN ISO 7730, in *Proceedings of Moving Thermal Comfort Standards into 21stCentury*, UK, pp. 31–44 (2001)
24. G. Alfano, F.R. d'Ambrosio, G. Riccio, Sensibility of the PMV index to variations of its independent variables, in *Proceedings of Thermal Comfort Standards into the 21stCentury*, Windsor, pp. 158–165 (2001)
25. E. Arens, M.A. Humphreys, R. de Dear, H. Zhang, Are 'class A' temperature requirements realistic or desirable? Build. Environ. **45**, 4–10 (2009)
26. J.F. Nicol, M.A. Humphreys, Maximum temperatures in European office buildings to avoid heat discomfort. Sol. Energy **81**(3), 295–304 (2007)
27. D. Robinson, F. Haldi, Model to predict overheating risk based on an electrical capacitor analogy. Energy Build. **40**, 1240–1245 (2007)
28. J.F. Nicol, M. Wilson, A critique of European standard EN 15251: strengths, weaknesses and lesson for future standards. Build. Res. Inf. **39**(2), 183–193 (2011)

Chapter 2
Comparison of the Ranking Capabilities of the Long-Term Discomfort Indices

Abstract The scientific literature and some standards offer a number of methods for the long-term evaluation of the general thermal comfort conditions in buildings and indices for predicting the likelihood of summer overheating in the indoor environment. Such metrics might be useful tools for operational assessment of thermal comfort in existing buildings, for driving the optimization process of a new building, or for optimizing the operation of building systems. Focusing exclusively on the summer period, 16 long-term discomfort indices have been applied for assessing 54 different variants of a large office building. Such 54 building variants were obtained by combining different performance levels of four key-parameters of the building envelope: (i) insulation and airtightness (ii) solar factor of glazing units (iii) exposed thermal mass and (iv) night natural ventilation strategies. The calculated values of the 16 indices were compared in order to identify their similarities and differences. The indices returned different results in ranking the 54 variants and hence, when used to drive an optimization process of a building design, they would identify different building variants as the optimal ones. Suggestions for improvement of the indices and their use are drawn from this analysis.

2.1 Introduction

The scientific literature and the standards ISO 7730 [1] and EN 15251 [2] offer some methods for the long-term evaluation of the general thermal comfort conditions and for predicting the likelihood of summer overheating occurrences in a building. Such methods—usually indices that cumulate, in a variety of ways, over time and space a chosen short-term and local discomfort index—are used for comparing the effect on the indoor environment of alternative design strategies. For example, EN 15251 guides designers towards a two-step optimization procedure, which is based on the sequential use of two long-term discomfort indices,

S. Carlucci, *Thermal Comfort Assessment of Buildings*,
PoliMI SpringerBriefs, DOI: 10.1007/978-88-470-5238-3_2,
© The Author(s) 2013

the former based on the European adaptive comfort model and the latter on the Fanger model:

> The temperature limits presented in A.2 [*A/N: based on adaptive model*] should be used for the dimensioning of passive means to prevent overheating in summer conditions e.g. dimensions and the orientation of windows, dimensions of solar shading and the thermal capacity of the building's construction. Where the adaptive temperature limits presented in A.2 (upper limits) cannot be guaranteed by passive means mechanical cooling is unavoidable. In such cases the design criteria for buildings WITH mechanical cooling should be used [*A/N: based on Fanger model*] [2].

Pagliano and Zangheri [3] show that employing the long-term discomfort indices proposed by EN 15251 during this optimization procedure brings to discontinuities in the design. Other authors have used several long-term discomfort indices to analyze thermal comfort performance or to optimize buildings. Pane [4] measures the frequency of exceedance of both the threshold temperatures of 25 and 27 °C for studying the relationship between thermal mass and summer overheating. Schnieders [5] uses the frequency of overheating events for comparing several options of the glazing units and for assessing summertime climate in a passive house; the frequency of exceedance of the temperatures of 25 and 26 °C has been included in the software PHPP [6], which is used for the design and validation of the German *Passivhaus*. Nicol et al. [7] introduce an index called Overheating risk, and use it to assess the likelihood of the overheating phenomenon in a building during a 25-day analysis. Robinson and Haldi [8] also propose an index to predict the summer overheating risk caused by the charge and discharge of heat stimuli in an individual, based on an electrical capacitor analogy. Grignon-Massé et al. [9] use the index called Percentage outside range to assess the cooling performance of several building envelope retrofit techniques and ventilation strategies in offices and commercial buildings. Borgeson and Brager [10] propose two indices called Exceedance$_{PPD}$ and Exceedance$_{Adaptive}$ and use them to assess, through simulations, summer thermal discomfort caused in a reference free-floating building located in the 16 different climatic zones of California. Di Perna et al. [11] use Percentage outside range based on the European adaptive comfort model to assess the summer reduction of thermal discomfort offered by an increase of thermal mass. In the last years, long-term discomfort metrics are used more and more often in mathematical optimization of buildings where thermal comfort is set as the objective function of the optimization problem [12–23], or like a constraint or penalty function [24–28]. Wang and Jin [13] use a sum weighted method to solve a multi-objective optimization problem, and one of the terms is thermal discomfort defined as the square of the hourly-simulated Predicted mean vote (PMV). Kolokotsa et al. [24] and Mossolly et al. [20] instead use the square of the difference between a threshold PMV set by the user and the hourly-simulated PMV. Nassif et al. [16], Nassif et al. [17] and Kummert and André [18] minimize hourly-simulated Predicted percentage of dissatisfied (PPD) when optimizing HVAC control system strategy. Magnier and Haghighat [25] build an objective function multiplying the average PMV over the whole year and over all occupied zones, by a function directly proportional to the number of hours when the absolute value of PMV is higher than 0.5. Corbin et al. [21] use, as an objective function, the offset of actual PMV with respect to neutrality weighted with the floor area of every zone of the building. Emmerich et al. [19] assess long-term thermal comfort in a building by counting the frequency of hourly exceedance of a threshold temperature fixed at 28 °C, and optimize a building by minimizing such long-term discomfort metric. Loonen et al. [22] use the same strategy, but choose a temperature threshold fixed at 25 °C. Hoes et al. [26, 28] minimize summer overheating and winter underheating hours, and set a constraint on the maximum number of discomfort hours—fixed at 200 h—in order to ensure a minimum thermal comfort level. Angelotti et al. [14] use a long-term index based on PMV to optimize the design of ground exchangers and night ventilation strategies. Stephan et al. [23] used the Percentage outside range and the Degree-hour criterion to optimize openings for night natural ventilation in order to activate the thermal mass and hence reduce thermal discomfort.

After having reviewed a number of long-term discomfort indices in Chap. 1, a comparison of their ranking capabilities is presented in this chapter.

2.2 The Adopted Methodology

The indices for the long-term evaluation of the general thermal comfort conditions in buildings might be useful tools for driving the optimization process of a new building, for optimizing the operation of building systems, or for the operational assessment of thermal comfort conditions in existing buildings. The methodology used here to understand their similarities and differences and their strengths and limitations consists of four phases.

In Phase 1, a number of variants of a reference building are generated by varying four key-parameters: (i) envelope resistance to heat flows and air infiltration (ii) solar factor of glazing unit of the windows (iii) exposed internal thermal mass, and (iv) operable window area for night natural ventilation. Various levels of performance are assumed for each key-parameter, and a summary is reported in Table 2.1. The focus is on the building envelope and passive strategies since the procedure proposed in EN 15251 has to be applied in two steps (first optimize envelope and passive systems towards an adaptive-based long-term discomfort index, and if comfort conditions cannot be met, optimize envelope and active systems towards a Fanger-based long-term discomfort index), and it is likely to produce discontinuities in case that the two indices will not find as optimal the same variants concerning envelope [3].

Then, 54 building variants are constructed by combining the various levels of performance of the four key-parameters (Table 2.2).

In Phase 2, all 54 building variants are simulated with EnergyPlus [29] and the hourly indoor operative temperature, humidity, air velocity and PMV for each zone of the building are stored in a database.

Table 2.1 Summary of variations for each key-parameter

Key features	Id of the parameter	Reference for the parameter	Building component or strategy	Value	Unit of measure
Envelope quality (U-value and air tightness)	0	Dlgs 311 [30]	Roof (Rome)	0.36	$W\ m^{-2}K^{-1}$
			Wall (Rome)	0.32	$W\ m^{-2}K^{-1}$
			Floor (Rome)	0.36	$W\ m^{-2}K^{-1}$
			Window (Rome)	2.40	$W\ m^{-2}K^{-1}$
			Air permeability	5.0	$m^3\ h^{-1}m^{-2}$
	+	SIA 380/1 [57]	Roof	0.20	$W\ m^{-2}K^{-1}$
			Wall	0.20	$W\ m^{-2}K^{-1}$
			Floor	0.20	$W\ m^{-2}K^{-1}$
			Window	1.20	$W\ m^{-2}K^{-1}$
			Air permeability	0.5	$m^3\ h^{-1}m^{-2}$
Solar protection (Solar factor)	–	Existing typical	Façade N	–	–
			Façade NE-NW	0.70	–
			Façade E-SE-S-SW-W	0.70	–
	0	Medium	Façade N	–	–
			Façade NE-NW	0.40	–
			Façade E-SE-S-SW-W	0.40	–
	+	SIA 380/1 [57]	Façade N	–	–
			Façade NE-NW	0.27	–
			Façade E-SE-S-SW-W	0.15	–

(continued)

Table 2.1 (continued)

Key features	Id of the parameter	Reference for the parameter	Building component or strategy	Value	Unit of measure
Thermal mass	−	Low internal thermal mass	External wall	4.0[a]	Wh m^{-2}K^{-1}
			Ceiling	11.0[a]	Wh m^{-2}K^{-1}
			Floor	4.1[a]	Wh m^{-2}K^{-1}
			Internal wall	2.3[a]	Wh m^{-2}K^{-1}
			Whole building	20.0[b]	Wh m^{-2}K^{-1}
	0	Medium internal thermal mass	External wall	15.4[a]	Wh m^{-2}K^{-1}
			Ceiling	18.6[a]	Wh m^{-2}K^{-1}
			Floor	12.7[a]	Wh m^{-2}K^{-1}
			Internal wall	8.9[a]	Wh m^{-2}K^{-1}
			Whole building	50.0[b]	Wh m^{-2}K^{-1}
	+	High internal thermal mass	External wall	15.4[a]	Wh m^{-2}K^{-1}
			Ceiling	22.1[a]	Wh m^{-2}K^{-1}
			Floor	22.4[a]	Wh m^{-2}K^{-1}
			Internal wall	18.8[a]	Wh m^{-2}K^{-1}
			Whole building	80.0[b]	Wh m^{-2}K^{-1}
Natural ventilation	−	No ventilation	% of operable window area	0	%
	0	Medium ventilation	% of operable window area	25	%
	+	Large ventilation	% of operable window area	50	%

[a] Thermal capacity is calculated according to [32] and expressed per area of building component
[b] Thermal capacity is calculated according to [31] and expressed per zone net floor area

Table 2.2 Building variants obtained from the combination of different options of the four key-parameters

Name of the variant	Thermal resistance	Air tightness	Thermal mass	Solar protection	Night natural ventilation
Variant 01	0	0	–	+	–
Variant 02	0	0	–	+	0
Variant 03	0	0	–	+	+
Variant 04	0	0	–	0	–
Variant 05	0	0	–	0	0
Variant 06	0	0	–	0	+
Variant 07	0	0	–	–	–
Variant 08	0	0	–	–	0
Variant 09	0	0	–	–	+
Variant 10	0	0	0	+	–
Variant 11	0	0	0	+	0
Variant 12	0	0	0	+	+
Variant 13	0	0	0	0	–
Variant 14	0	0	0	0	0
Variant 15	0	0	0	0	+
Variant 16	0	0	0	–	–
Variant 17	0	0	0	–	0
Variant 18	0	0	0	–	+
Variant 19	0	0	+	+	–
Variant 20	0	0	+	+	0
Variant 21	0	0	+	+	+
Variant 22	0	0	+	0	–
Variant 23	0	0	+	0	0
Variant 24	0	0	+	0	+
Variant 25	0	0	+	–	–
Variant 26	0	0	+	–	0
Variant 27	0	0	+	–	+
Variant 28	+	+	–	+	–
Variant 29	+	+	–	+	0
Variant 30	+	+	–	+	+
Variant 31	+	+	–	0	–
Variant 32	+	+	–	0	0
Variant 33	+	+	–	0	+
Variant 34	+	+	–	–	–
Variant 35	+	+	–	–	0
Variant 36	+	+	–	–	+
Variant 37	+	+	0	+	–
Variant 38	+	+	0	+	0
Variant 39	+	+	0	+	+
Variant 40	+	+	0	0	–

(continued)

Table 2.2 (continued)

Name of the variant	Thermal resistance	Air tightness	Thermal mass	Solar protection	Night natural ventilation
Variant 41	+	+	0	0	0
Variant 42	+	+	0	0	+
Variant 43	+	+	0	–	–
Variant 44	+	+	0	–	0
Variant 45	+	+	0	–	+
Variant 46	+	+	+	+	–
Variant 47	+	+	+	+	0
Variant 48	+	+	+	+	+
Variant 49	+	+	+	0	–
Variant 50	+	+	+	0	0
Variant 51	+	+	+	0	+
Variant 52	+	+	+	–	–
Variant 53	+	+	+	–	0
Variant 54	+	+	+	–	+

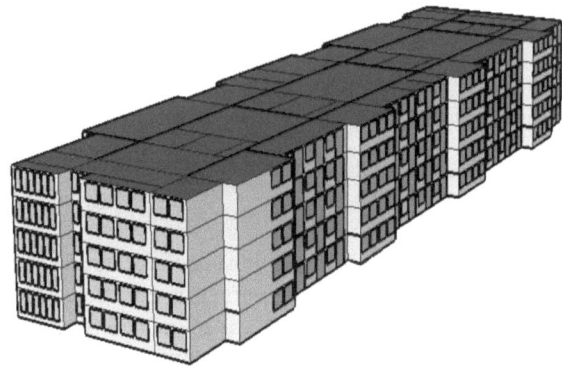

Fig. 2.1 Tridimensional visualization of the building model

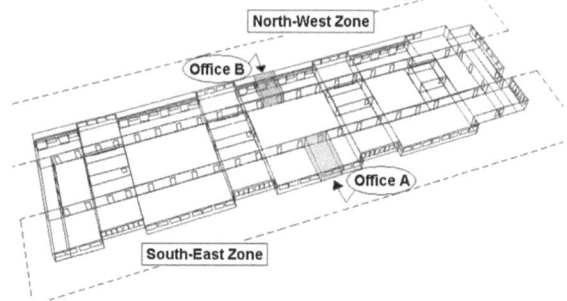

Fig. 2.2 Visualization of zoning of the standard floor

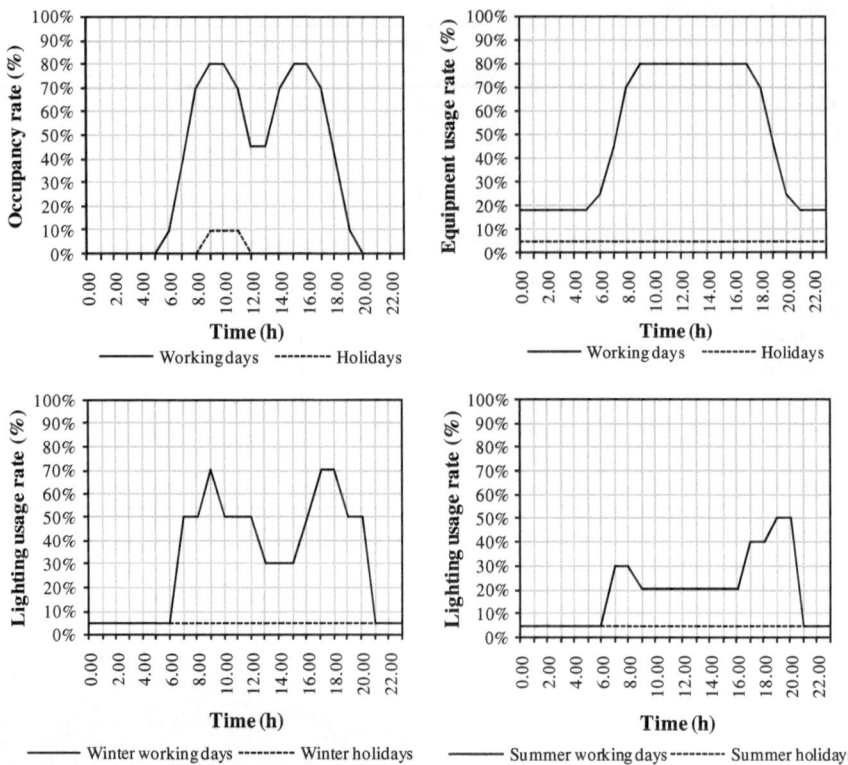

Fig. 2.3 Time schedules of the occupancy rate and the lighting and office equipment usage rates used in the simulations

In Phase 3, a statistical analysis of the hourly indoor operative temperatures in each zone of each building variant is conducted and the 16 long-term discomfort indices reviewed in Chap. 1 are calculated for each building variant.

In Phase 4, all long-term discomfort indices are analyzed through the comparison of the results in subsets; their similarities and differences are identified in order to obtain information on the implications of the choice of the index on the design process.

2.2.1 The Reference Building Model

The reference building is a large five-story office building (Fig. 2.1), characterized by an occupied volume of 32 706 m³ and an external surface of 8 501 m²; thus its shape ratio (S/V) is 0.26 m⁻¹. The value of the ratio between window area and total facade area is 40 %.

The standard floor, which is repeated for each story, was zoned in five thermal zones: southeast zone (20 offices similar to office A, 710 m^2), northwest zone (21 offices similar to office B, 514 m^2), northeast zone (3 offices, 66 m^2), southwest zone (3 offices, 33 m^2), core zone (corridors, bath and stair-lift, 935 m^2). The geometry and the thermal zoning of the standard floor are shown in Fig. 2.2.

The technical features of the reference building are in agreement with the requirements of the Italian Law Decreto Legislativo n. 311 (DLgs 311) [30], valid for new buildings built from 2010 onward, specifically for a building located in Rome (IT). Steady-state thermal transmittance, or U-value, of external walls is equal to 0.36 Wm^{-2}K^{-1}, U-value of the roof is equal to 0.32 Wm^{-2}K^{-1}, U-value of the basement floor is equal to 0.36 Wm^{-2}K^{-1} and U-value of windows is equal to 2.40 Wm^{-2}K^{-1}. The double-glazing units are filled with air and are characterized by a g-value of 0.70; moreover, external solar shading does not protect them. Specified thermal capacity[1] of the building is 50 Wh m^{-2} K^{-1} and air permeability[2] of the envelope, defined according to SIA 180 [31] is 5.0 m^3 h^{-1}m^{-2}. Internal loads follow the typical daily schedules shown in Fig. 2.3.

During summer nights, the windows are open to provide night ventilation from 8:00 PM to 7:00 AM whether the hourly outside air temperature is lower than the hourly indoor air temperature.

2.2.2 Physical Models Set in the Numerical Model

The energy simulations of the building were run with the software EnergyPlus version 6.0.0.23. It is validated against the standard ANSI/ASHRAE 140 [33]. The building models were set so as to carefully reproduce the physical phenomena that affect the thermal behavior of the building, although this caused an increase of the computational time. The settings tuned differently from the EnergyPlus defaults were: calculation of the sun position, calculation of heat conduction in solids, calculation of natural convection heat transfer coefficients and calculation of both voluntary (ventilation) and involuntary (infiltrations) air flows.

2.2.2.1 Calculation of the Sun Position

The weather file contains information about the intensity of the available solar radiation in a certain location. The simulation engine has to calculate the sun paths to determine the most reliable amount of solar irradiation incident on any surface

---

[1] Thermal capacity for a building or a zone of a building is calculated according to the SIA definition [31] based on EN ISO 13786 [32]. It is expressed per zone net floor area.

[2] Air permeability is calculated according to [31] and it is referred to the area of the envelope.

of the building model in a specified period of the year. By default, EnergyPlus
updates the sun paths every 20 days. However, since the simulations aim at
identifying the effect also of the solar factor of glazing on the overall building
performance, the updating period of the sun position was reduced to 7 days.

Shadow patterns on exterior surfaces (walls, windows and roofs) caused by
detached shading, overhangs, side-fins and exterior surfaces of all zones of the
model are computed. All beam solar radiation entering a zone is assumed to fall on
the floor, where it is absorbed according to solar absorbance of its finishing layer.
Radiation reflected by floor is added to the transmitted diffuse radiation, which is
assumed to be uniformly distributed on all interior surfaces.[3]

2.2.2.2 Heat Conduction Method

The thermal mass of opaque components of the building envelope is also a design
parameter of the analysis. It was varied in three levels: low ($-$), medium (0) and
high (+). In order to accurately model opaque envelope components with high
inertia, the default model for calculating the conduction transfer function was not
suitable [34], thus it was replaced by the finite difference method. This setting
requested to decrease the duration of the time step from 10 min (Six time steps in
one hour by default) to 3 min (20 time steps in one hour) [35].

2.2.2.3 Natural Convection Heat Transfer Coefficient

The adaptive-surface-convection algorithms were chosen to calculate natural
convection heat exchange near the external and internal surfaces; thus, EnergyPlus
dynamically selects the algorithm that best applies among a number of different
available convection models. The models for calculating the natural-convection-
heat-transfer coefficient that are currently implemented in EnergyPlus, allow to set
a fixed value of the coefficient or to calculate it as a function of (i) HVAC system
type (ii) HVAC operating status (iii) air change per hour (iv) surface classification—floor, wall, roof, window—(v) range of tilt angle (vi) stability of convective
flows.

For the natural buoyancy airflow regime on inside surfaces, Fohanno's and
Polidori's algorithm [36] is used for vertical walls. Alamdari's and Hammond's
algorithm [37] is used for stable and unstable horizontal surfaces. A horizontal
surface is stable when the heat flux fosters airflow to move away from the surface;
on the contrary, it is unstable when the heat flux retards airflow to move away from
the surface. Walton's algorithm [38] is also used for both stable and unstable tilted

---

[3] The *FullInteriorAndExteriorWithReflections* option was not used since not all zones of the
building were convex. The convexity of the zone is a necessary condition for the calculation of
the amount of energy falling on every indoor surface of the zone.

surfaces. The ISO 15099 algorithm [39] is used for windows. The natural convection heat transfer coefficient is a function of the conductivity of air, the height of the window and the Nusselt number. The Nusselt number is calculated from the Rayleigh number based on the height of the window. It also depends on the value of the tilt angle created by a window with the vertical direction, thought several empirical correlations.

Thermal properties of air are evaluated at the mean film temperature ($\theta_{m,f}$) that is updated at each time step. Thermal conductivity of air is calculated with the equation $\lambda = 2.873 \cdot 10^{-3} + 7.76 \cdot 10^{-8} \, \theta_{m,f}$ whilst kinematic viscosity of air with the formula $\mu = 3.723 \cdot 10^{-6} + 4.94 \cdot 10^{-8} \, \theta_{m,f}$.

All models for modeling the natural buoyancy regime, currently implemented in EnergyPlus, exclusively relate convective heat transfer at internal room surfaces with the sole temperature difference between the undisturbed air temperature (the temperature of the node representing the thermal zone) and the surface temperature (it is assumed one uniform temperature for the whole surface), however, it locally depends on the indoor air temperature distribution and air velocity near the surfaces [40].

2.2.2.4 Infiltrations and Ventilation

In order to model the voluntary airflow through windows and doors and involuntary airflow caused by envelope leakage, the AirflowNetwork module [41–43] integrated in EnergyPlus was used. It allows simulating multi-zone airflows driven by wind, stack effect and forced ventilation whether present, through envelope leakage and indoor and outdoor windows and doors.

The AirflowNetwork method does not consider the effect of the air thermal capacitance and consists in creating a node for each zone of the building and in connecting them through linkages. Such model consists of three sequential steps that are executed for each node of the model:

1. Calculation of air pressure and airflow rate.
2. Calculation of temperature and humidity.
3. Calculation of sensible and latent load.

The pressure difference across a linkage component between the nodes n and m ($\Delta P_{n,m}$) is assumed to be governed by the Bernoulli's equation, and it is given by $\Delta P_{n,m} = (P_n - P_m) + \Delta P_s + \Delta P_w$, where P_n and P_m are the total pressure respectively at node n and m, ΔP_s is the pressure difference due to density and height differences (stack effect) and ΔP_w is the pressure difference due to the wind.

Based on the total node pressures, the model calculates airflow through each leakage based on the pressure versus airflow relationship defined for each component. Then, using such airflows and the piece of information that describes the HVAC components implemented, the node temperature and the humidity ratio are calculated for each node.

The sensible and latent loads obtained in this step are then used in the heat and moisture balance equations of each zone to predict HVAC system loads and to calculate the final air temperatures, humidity ratios and pressures of each zone.

2.2.2.5 Cracks

The AirflowNetwork module uses Eq. 2.1 to calculate the airflow through a crack in a surface. The airflow rate, Q, in kg s^{-1}, is given by:

$$Q = C_T \cdot C_Q \cdot \Delta P^n_{crack} \qquad (2.1)$$

where C_T is the airflow rate coefficient (kg s^{-1}Pa^{-n} @ 1 Pa), C_Q is the temperature correction factor with respect to the reference condition (dimensionless), ΔP_{crack} is the pressure difference across the crack (Pa), n is the airflow exponent (dimensionless). In particular, the valid range for the value of the airflow exponent is from 0.5 to 1.0, with the default value of 0.65.

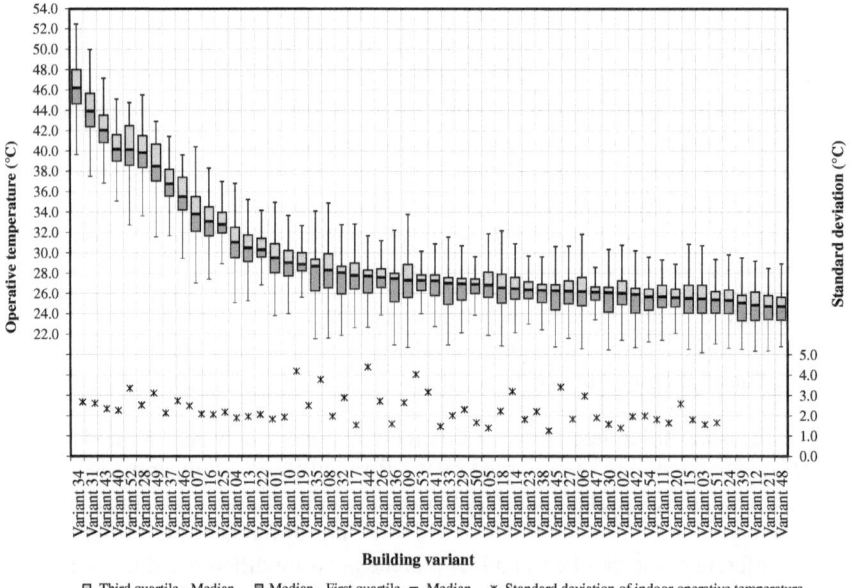

Fig. 2.4 Indoor operative temperatures and standard deviation, both weighted for the occupation hours and the net volume of each zone, for a large office building in free-floating mode from 15 May to 30 September

Table 2.3 Features of the best five building variants

Name of the variant	Thermal resistance	Air tightness	Thermal mass	Solar protection	Night natural ventilation
Variant 48	+	+	+	+	+
Variant 21	0	0	+	+	+
Variant 12	0	0	0	+	+
Variant 39	+	+	0	+	+
Variant 24	0	0	+	0	+

2.2.3 Variations of the Technical Features of the Building Envelope

Optimization of the building was executed thought the combination of four key-parameters (envelope quality, solar factor of glazing, thermal mass of opaque components, natural ventilation). The variations for each key-parameter are summarized in Table 2.1.

2.2.4 Building Variants

54 building variants were derived combining every option of the four key-parameters of Table 2.1. All combinations are summarized in Table 2.2.

2.2.5 Comparison of the Building Variants

In order to compare the results of the simulation runs, the 54 variants have been ranked according to the average (over time and space) indoor operative temperature, from the highest to the lowest. For every variant, the hourly operative temperature was obtained via dynamic simulation for each zone of the model and they were averaged weighting by the net volume of each zone. Then, a statistical description of the time series of the obtained hourly space averaged indoor operative temperatures was performed to identify the median, the first quartiles and the third quartiles. A box-and-whisker-plot representation is used to visualize the result of this statistical description of the space averaged hourly operative temperatures during occupation hours in the multi-zone model for every building variant (Fig. 2.4).

For each variant, the box represents the distance between the first and third quartiles of the time series of the space averaged operative temperatures, and the solid horizontal segment is the median. The whiskers (indicated through error bars) represent the tails of the distribution and show the highest and lowest data points (space averaged maximum or minimum hourly operative temperatures) or otherwise 1.5 times the box range [44].

The simulation period is from 15 May to 30 September corresponding to a typical summer in Italy.

Figure 2.4 shows that the variants characterized by the lowest indoor operative temperatures have also low average standard deviation of the indoor operative temperature (excepted for Variant 39). In other words, this means that those variants characterized by the lowest average indoor operative temperatures such as Variants 48, 21, 12 and 39, also show a low fluctuation of indoor operative temperature. Those variants are characterized by low value of the solar factor (between 0.15 and 0.27), high night ventilation rates and relatively high thermal mass, as shown in Table 2.3 that reports the features of the five variants with the lowest average indoor operative temperatures.

Their comparison suggests that, in order to reduce indoor operative temperature during summer:

- High insulation levels and reduced airtightness have to be coupled with effective strategies for discharging internal energy.
- Effective strategies for discharging internal energy are essential—natural night ventilation is set to the highest value (50 % of the window area is assumed to be open during nighttime) for all seventh best variants.
- An effective reduction of solar gains is necessary (solar factor lower than 0.15 for east/southeast/southwest/west facades and lower than 0.27 for northeast/northwest facades)—four of the best five variants require the lowest value of the solar factor.
- High exposed thermal mass (80 Wh $K^{-1}m^{-2}$) is desirable: if the other features are held constant, increasing thermal mass, contributes to lower indoor operative temperature. We can compare:

  - Variant 03, Variant 12 and Variant 21: the thermal mass value passes from 20 to 50 up to 80 Wh $m^{-2}K^{-1}$ respectively.
  - Variant 39 and Variant 48: the thermal mass value passes from 50 to 80 Wh $m^{-2}K^{-1}$ respectively.

2.3 Discussion of the Results

Sixteen long-term indices were calculated using as input the indoor environmental conditions simulated for each building variant (i.e., the same input for all 16 indices):

1. Percentage of occupied hours when actual PMV is outside the Fanger comfort range, $POR_{Fanger,PMV}$ [1, 2].

Table 2.4 Results of the Pearson correlation analysis between each index for the long-term evaluation and average indoor operative temperature (all indices are statistically significant, $p < 0.05$)

Name of the index	Pearson correlation coefficient, r	Strength of association
$DhC_{Fanger, ISO\ 7730}$	1.00	Strong
$DhC_{Fanger, EN\ 15251}$	1.00	Strong
$DhC_{Adaptive}$	0.98	Strong
NaOR	0.98	Strong
<PPD>	0.95	Strong
Sum_PPD	0.95	Strong
$PPDwC_{ISO\ 7730}$	0.95	Strong
$PPDwC_{EN\ 15251}$	0.95	Strong
$POR_{Adaptive}$	0.90	Strong
$CISBE_A$	0.88	Strong
$Exceedance_{Adaptive}$	0.85	Strong
$Exceedance_{PPD}$	0.84	Strong
RHOR	0.83	Strong
$POR_{Fanger,\theta op}$	0.82	Strong
$POR_{Fanger,PMV}$	0.81	Strong
$CISBE_J$	0.69	Strong

2. Percentage of occupied hours when the indoor operative temperature[4] is outside the Fanger comfort range, $POR_{Fanger,\theta_{op}}$ [1, 2].
3. Percentage of occupied hours when the indoor operative temperature is outside the EN adaptive comfort range, $POR_{Adaptive}$ [2].
4. Degree-hour criterion based on the Fanger comfort model as proposed in the ISO 7730 version, $DhC_{Fanger, ISO\ 7730}$ [1].
5. Degree-hour criterion based on the Fanger comfort model as proposed in the EN 15251 version, $DhC_{Fanger, EN\ 15251}$ [2].
6. Degree-hour criterion based on the EN adaptive comfort model, $DhC_{Adaptive}$ [2].
7. PPD-weighted criterion as stated in ISO 7730, $PPDwC_{ISO\ 7730}$ [1].
8. PPD-weighted criterion as stated in EN 15251 version, $PPDwC_{EN\ 15251}$ [2].
9. Average PPD, <PPD> [1].
10. Accumulated PPD, Sum_PPD [1].
11. CISBE Guide A criterion, $CISBE_A$ [45].
12. CISBE Guide J criterion, $CISBE_J$ [46].
13. $Exceedance_{PPD}$ [10].
14. $Exceedance_{Adaptive}$ [10].
15. Nicol et al.'s overheating risk, NaOR [7].
16. Robinson's and Haldi's overheating risk, RHOR [8].

---

[4] The limit values of PMV have been translated into operative temperatures adopting the assumptions contained in ISO 7730: metabolic rate of 1.2 met, clothing resistance of 0.5 clo, air velocity of 0.1 m s^{-1}, relative humidity of 50%, mean radiant temperature equal to air temperature equal to operative temperature. The fact that these variables are set to fixed values should be kept in mind when evaluating the index.

Table 2.5 Suggested values for interpreting the strength of linear dependence between two variables according to [48]

Strength of association	Pearson coefficient, r	
	Positive	Negative
None or very weak	$0.0 \div 0.1$	$-0.1 \div 0.0$
Weak	$0.1 \div 0.3$	$-0.3 \div -0.1$
Moderate	$0.3 \div 0.5$	$-0.5 \div -0.3$
Strong	$0.5 \div 1.0$	$-1.0 \div -0.5$

One of the results of the analysis is that these indices do not provide a similar ranking of the thermal comfort performance of the building variants. Their trends over the same sample of buildings (the 54 variants) are different in the absolute value and shape: some remain almost stable at high values while others indicate that the likelihood of thermal discomfort is almost zero. Obviously the values of the indices we present in this book are specific for the described case study since they depend on the reference building that has been chosen, the assumptions made about the occupant-related input variables, the climate for which it has been simulated, the chosen calculation period, etc.; but the fact that they provide quite different rankings of the variants and hence drive the design process towards different *optimal* variants is unlikely to depend on the specific case study. At the minimum, we have identified at least one case where the various indices would deliver different conclusions to the designer.

Average indoor operative temperature, being the single variable with more weight in the determining comfort in moderate environments [47] is used here for an overall assessment of all indices. They are assessed using the Pearson correlation analysis in order to identify the strength of linear dependence between the trends of each index with the trend of the average indoor operative temperature (Table 2.4). Resulting values of the Pearson correlation coefficient, r, are interpreted referring to Buda and Jarynowski [48] (Table 2.5), even if such criteria have to be intended as a guideline and not in a too strict way [49].

- All indices show a strong correlation with the average indoor operative temperature: some are close to a perfect linear dependence whilst others are closer to a moderate association.
- The Degree-hour criteria are close to a perfect linear dependence with the average indoor operative temperature. This can be related to their dependence by the offset between the indoor operative temperature and a certain boundary temperature. Decreasing the indoor operative temperature decrease the exceedance outside the boundary.
- Also NaOR is strongly associated and also it depends on the offset between the indoor operative temperature and the theoretical comfort temperature calculated according the EN adaptive comfort model; RHOR is less correlated likely because it aims at measuring the heat stimulus due to the exceedance above and below 25 °C.

- Both Sum_PPD and <PPD> feature the same Pearson correlation coefficient. Even if these two indices are characterized by different units of measure, they are similar in the meaning: the former exclusively sums the hourly PPDs in a specified occupied period whilst the latter divides the resulting accumulated value, calculated as before, by the number of occupied hours.
- Also the two PPD-weighted criteria reach the same Pearson correlation coefficient. Their formats exclusively differ because one counts as uncomfortable also those occupied hours when the hourly PMV is equal to the boundary values while the other does not; however, the hourly-simulated PMV was never exactly equal to 0.5 during all simulations.
- On the contrary, the three Percentage outside range indices are characterized by different levels of association. This depends by a diverse comfort model (Fanger and EN adaptive) and, regarding the sole indices based on the Fanger model, the metric used for expressing the boundary (PMV or operative temperature) does not affect significantly the results of the correlation analysis.
- The two Exceedance indices do not differ significantly, although they identify discomfort occurrences using different comfort models.
- The two CIBSE criteria show a significant difference—r (CIBSE$_J$) = 0.69; r (CIBSE$_A$) = 0.88. The simple change of the base temperature (respectively 25 and 26/28 °C[5]) has a significant impact on the relationship between the frequency of exceedance of the fixed temperature and indoor thermal conditions.

Then, the indices were compared for subcategories for highlighting trends, similarities and differences. The most general classification of indices for long-term evaluation is based on their output: a percentage value, and an accumulation of hours (usually weighted by a weighting factor).

2.3.1 Indices Unrepresentable on the Percentage Scale

The long-term discomfort indices belonging to this category are:

1. Degree-hour criterion based on the Fanger comfort model as proposed in ISO 7730, DhC$_{Fanger, ISO 7730}$.
2. Degree-hour criterion based on the Fanger comfort model as proposed in EN 15251, DhC$_{Fanger, EN 15251}$.
3. Degree-hour criterion based on the EN adaptive comfort model, DhC$_{Adaptive}$.
4. *PPD*-weighted criterion as stated in ISO 7730, PPDwC$_{ISO 7730}$.
5. *PPD*-weighted criterion as stated in EN 15251, PPDwC$_{EN 15251}$.
6. Accumulated PPD, Sum_PPD.

They are weighted degree-hour indices, meaning that they accumulate all time intervals (generally one hour) when an uncomfortable condition occurs, each time

---

[5] 26 °C for the bedrooms and 28 °C for other rooms.

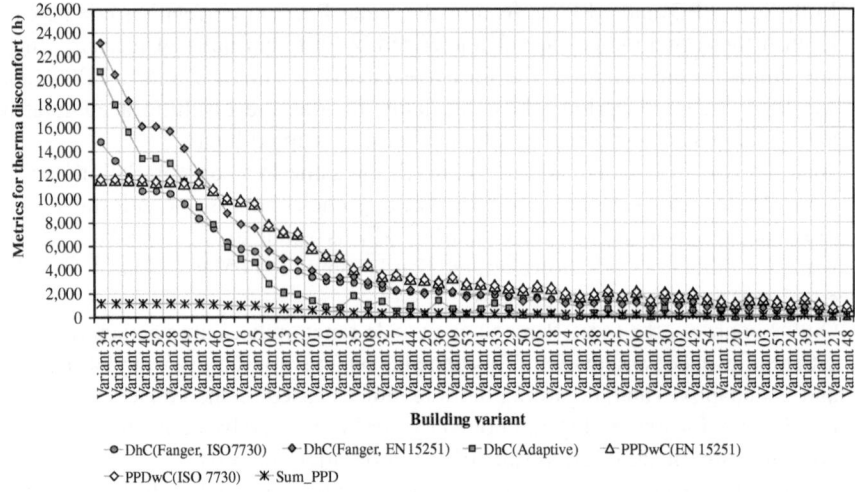

Fig. 2.5 Comparison of cumulative indices unrepresentable on a percentage scale

interval being previously multiplied for a certain weighting factor which could be the simple difference between hourly operative temperature and a boundary temperature of a specified comfort category (DhC), or the hourly PPD (Sum_PPD), or a more complex weighting (PPDwC). All these indices are based on a comfort model (Fanger or adaptive); DhC and PPDwC are also category-dependent and asymmetric[6] indices. On the contrary, Sum_PPD does not depend on comfort categories and is symmetric. Their behaviors are compared in Fig. 2.5.

Since DhCs indices are the accumulation of the temperature offset from the comfort threshold per each hour, they are higher for those building variants characterized by higher indoor operative temperatures. For this reason, the DhCs have trends similar in shape to the trend of average indoor operative temperature. PPDwCs (both the ISO and EN versions) accumulate, per each discomfort hour, the product of a weighing factor and the time span; this weighting factor corresponds to the ratio between the actual PPD and the boundary PPD of a specified category; hence the hourly weighting factor aims at a constant value for PPD close to 100 % and for this reason it shows a tendency to saturate to a finite value (corresponding to PPD close to 100 % in each hour of the calculation period). Sum_PPD accumulates the hourly value of PPD for each discomfort hour; hence it saturates to a finite value that corresponds to the product of PPD close to 100 % per the total number of hours of the calculation period (Fig. 2.6).

---

[6] That is they weigh in a different way hours of overheating and of overcooling in summer; for the complete definitions and more details on the features of the indices see Chap. 1. The issue whether and when to include summer overcooling in the indices (i.e., the choice between symmetric and asymmetric indices) is subject of debate in literature and in the Standardization Committees.

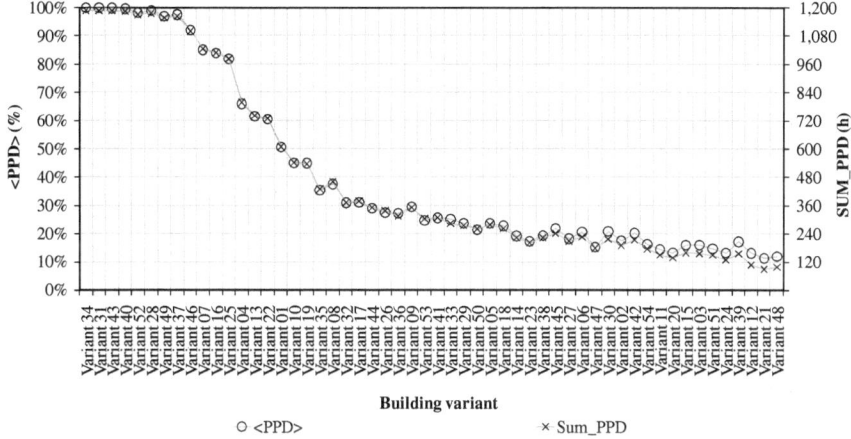

Fig. 2.6 Comparison between Sum_PPD and <PPD>

PPDwC and Sum_PPD show similar trends, but their values differ for an order of magnitude.

All these indices are closely correlated to the average indoor operative temperature (the Pearson correlation coefficient, r, ranges from 0.95 to 1.00). However they have the limitation that their scale does not have a upper bound; hence they are impractical for comparing buildings that differ in geometrical or physical structure, or time schedule, or external climate, etc. (Fig. 2.5).

Also, Exceedance$_{PPD}$ and Exceedance$_{Adaptive}$ are weighted degree-hour indices since they weight every hour, when actual PPD is higher than 20 %, with a weighting factor that is the actual or forecasted occupation rate, but the summation over the

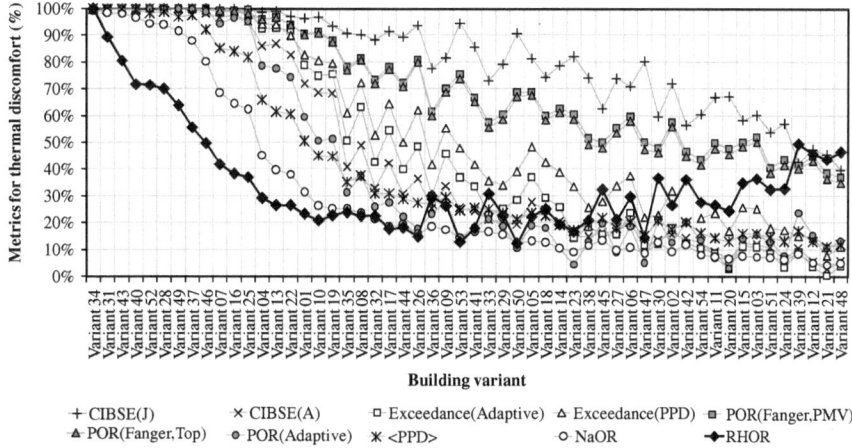

Fig. 2.7 Comparison of long-term discomfort indices representable on a percentage scale

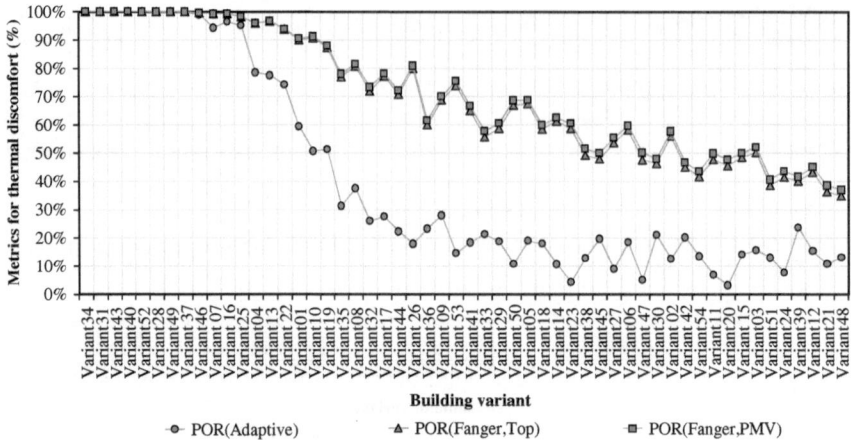

Fig. 2.8 Comparison of the trends when category boundaries are expressed via PMV or operative temperature

whole calculation period is divided by the summation of all hourly occupation rates, resulting in a percentage index; so they are discussed in the next paragraph.

2.3.2 Indices Representable on the Percentage Scale

They are a large and complex group, composed by indices based or not on comfort models, indices dependent or not by comfort categories, symmetric and asymmetric indices. Some of them are provided together with discomfort thresholds chosen somewhat arbitrarily rather than based on a comfort model. The long-term discomfort indices belonging to this category, and represented in Fig. 2.7, are:

1. Percentage of occupied hours when actual PMV is outside the Fanger comfort range, $POR_{Fanger,PMV}$.
2. Percentage of occupied hours when the indoor operative temperature[7] is outside the Fanger comfort range, $POR_{Fanger,\theta_{op}}$
3. Percentage of occupied hours when the indoor operative temperature is outside the EN adaptive comfort range, $POR_{Adaptive}$.
4. Average PPD, <PPD>.
5. CISBE Guide J criterion, $CIBSE_J$.
5. CISBE Guide A criterion, $CIBSE_A$.

---

[7] The limit values of PMV have been translated into operative temperatures adopting the assumptions contained in ISO 7730 for summer conditions: metabolic rate of 1.2 met, clothing resistance of 0.5 clo, air velocity of 0.1 m s^{-1}, relative humidity of 50%, mean radiant temperature equal to air temperature equal to operative temperature. The fact that some variables are forced to assume predefined fixed values should be kept in mind when evaluating the index.

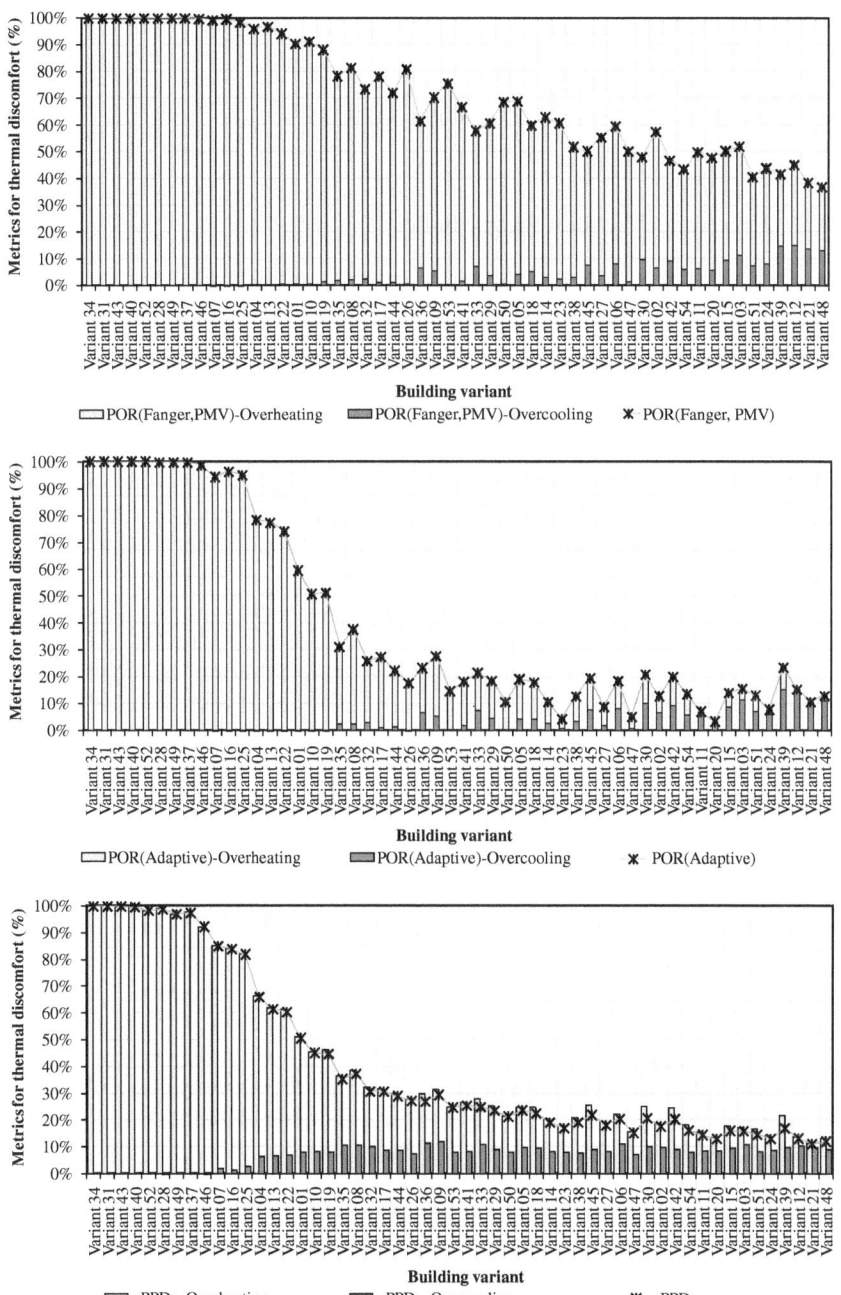

Fig. 2.9 Comparison between the global value of three symmetric indices and their component due to overheating and overcooling

7. Exceedance$_{PPD}$.
8. Exceedance$_{Adaptive}$.
9. Nicol et al.'s overheating risk, NaOR.
10. Robinson's and Haldi's overheating risk, RHOR.

2.3.2.1 Percentage Outside Range Indices

For POR$_{Fanger,PMV}$, hourly PMV is calculated using the simulated values of indoor dry-bulb air temperature, mean radiant temperature, indoor air relative humidity and assuming constant values for the metabolic activity set to 1.2 met, the clothing resistance set to 0.5 clo, air velocity set to 0.1 m s^{-1} and the external work set to zero met. During the occupied hours, whether hourly PMV is higher than +0.5 or lower than −0.5, an occurrence of discomfort is recorded.

For POR$_{Fanger,\theta op}$, the thermal neutrality operative temperature, results in 24.7 °C, assuming: air temperature equal to mean radiant temperature (and hence equal to operative temperature), indoor air relative humidity of 50 %, air velocity of 0.1 m s^{-1}, metabolic activity of 1.2 met, clothing resistance of 0.5 clo and external work of zero met. The Category II comfort range is identified by an offset from thermal neutrality temperature of ± 1.7 K, according to ISO 7730. As a result, the upper and lower boundary operative temperatures of the comfort range result respectively in 26.4 and 23.0 °C. An occurrence of discomfort is recorded during the occupied hours, when hourly indoor operative temperature falls outside the aforementioned comfort range. Compared with the previous index, also the relative humidity needs to be fixed a priori.

For POR$_{Adaptive}$, an occurrence of discomfort is recorded when the hourly indoor operative temperature falls outside the EU adaptive comfort range corresponding to Category II. In this case, the temperature range is defined by an offset of ±3 K from the adaptive theoretical comfort temperature. Such three indices are represented in Fig. 2.8.

POR$_{Fanger,PMV}$ and POR$_{Fanger, \theta op}$ show values that are very close—as one may expect—since they differ only because the latter is calculated by also imposing relative humidity equal to the fixed value of 50 % and the former by using the actual hourly values of relative humidity, and since humidity has only a little influence on thermal comfort temperature at moderate environments according to the Fanger comfort model [1]. While POR$_{Fanger,PMV}$ and POR$_{Fanger,\theta op}$ show (with some fluctuations) a progressive reduction of the discomfort percentage up to Variant 48; POR$_{Adaptive}$ reaches (with some fluctuations) the minimum discomfort value for Variant 20 and the index has a slightly increasing trend for those building variants with the lowest average indoor operative temperatures. Since, during hot periods, the boundary temperatures of the adaptive comfort range are higher than the boundary temperatures of the Fanger comfort range, the increase of discomfort might be due to the increase of hours when the indoor operative temperature is

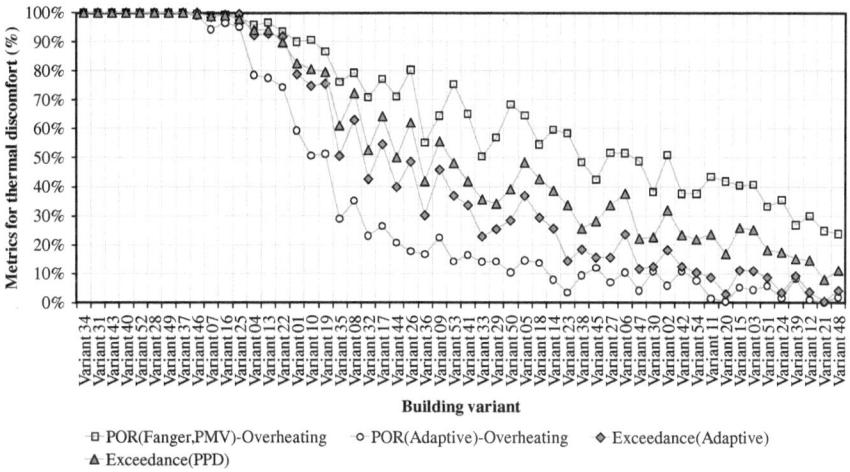

Fig. 2.10 Comparison among portions of $POR_{Fanger,PMV}$ and $POR_{Adaptive}$ due to sole overheating and Exceedance indices

lower than the adaptive comfort range (overcooling hours), but falls inside the Fanger range (comfort hours). We remind here that night hours are excluded from the calculation period in the presented case study.

2.3.2.2 Symmetric Indices and the Overcooling Phenomenon

In order to analyze if the detected overcooling phenomenon depends only on the definition of the categories, we compare the aforementioned $POR_{Fanger,PMV}$, $POR_{Adaptive}$, which are category-dependent indices, and the Average PPD (in short <PPD>), which is category-independent. We disaggregate the frequencies of exceedance above the comfort boundary (overheating occurrences) and below the comfort boundaries (overcooling occurrences) in Fig. 2.9.

$POR_{Fanger,PMV}$ records overcooling phenomenon, but discomfort due to over-heating is predominant for every variant. $POR_{Adaptive}$ records the overcooling phenomenon and it results the sole cause of discomfort for the three building variants with the lowest average operative temperatures. Also <PPD> records overcooling, and discomfort due to overcooling is comparable with overheating for those variants with the lowest average indoor operative temperatures.

High levels of natural night ventilation and high and medium level of thermal mass characterize the building variants with the lowest average operative temperature (described as affected by overcooling by these indices) (Table 2.3).

In synthesis (i) symmetric indices aim at detecting overcooling, which might take place, e.g., in variants characterized by high and medium level of thermal mass coupled with high levels of natural night ventilation, and this has a significant influence in the ranking of building variants; (ii) the difference between

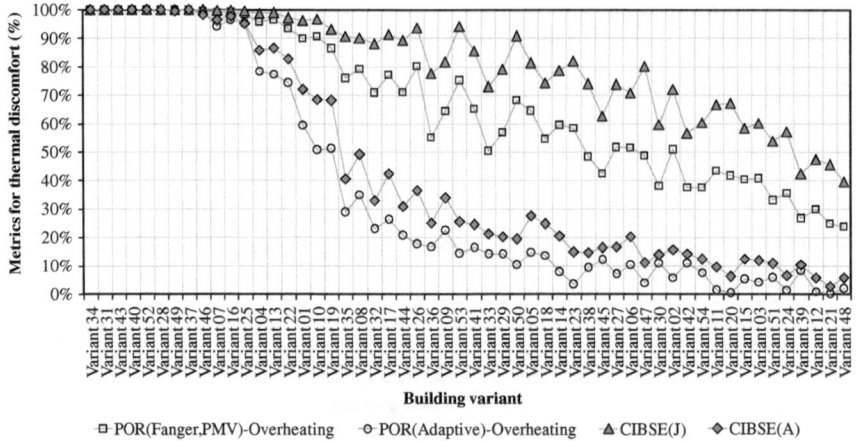

Fig. 2.11 Comparison between some of POR (only for the overheating part) and CIBSE criteria

POR$_{\text{Fanger,PMV}}$, POR$_{\text{Adaptive}}$ is due to the definition of the two comfort ranges in term of theoretical comfort temperature and extension in degrees; (iii) the Percentage outside range just measures for how many time intervals the indoor operative temperatures fall into a certain comfort range, on the contrary the <PPD> takes into account also the severity of discomfort in each hour since it is related to the likely human thermal perception measured by PPD.

From the above analysis, we derive that, in order to correctly interpret the results obtained by using symmetric indices, one should explicitly report their disaggregation in overheating and overcooling. In addition, since POR$_{\text{Fanger,PMV}}$ is a simple counter of the hours outside the comfort range and <PPD> provides a measure of human thermal perception by accounting for severity of discomfort in each hour, we believe that <PPD> should be preferred to POR$_{\text{Fanger,PMV}}$ when assessing thermal discomfort according to the Fanger comfort model.

2.3.2.3 Exceedance Indices

Exceedance$_{\text{PPD}}$ and Exceedance$_{\text{Adaptive}}$ are asymmetric indices, i.e., they take into account only discomfort due to overheating conditions in summer. For both indices discomfort is considered to occur when hourly indoor conditions exceed a discomfort threshold fixed at 20 %. It corresponds to a PPD value of 20 % (PMV = 0.85) for Exceedance$_{\text{PPD}}$ and to a temperature rise of 3.5 °C with respect to the optimal ASHRAE adaptive temperature for Exceedance$_{\text{Adaptive}}$.

The weighting factor used in both Exceedance$_{\text{PPD}}$ and Exceedance$_{\text{Adaptive}}$, is the ratio between hourly occupation rate and the accumulated occupation rate over the calculation period. If hourly occupation rate is almost the same in each zone of the building, Exceedance$_{\text{PPD}}$ and Exceedance$_{\text{Adaptive}}$ converge to the time frequency of exceedance of the specified comfort boundaries. In our case, since the

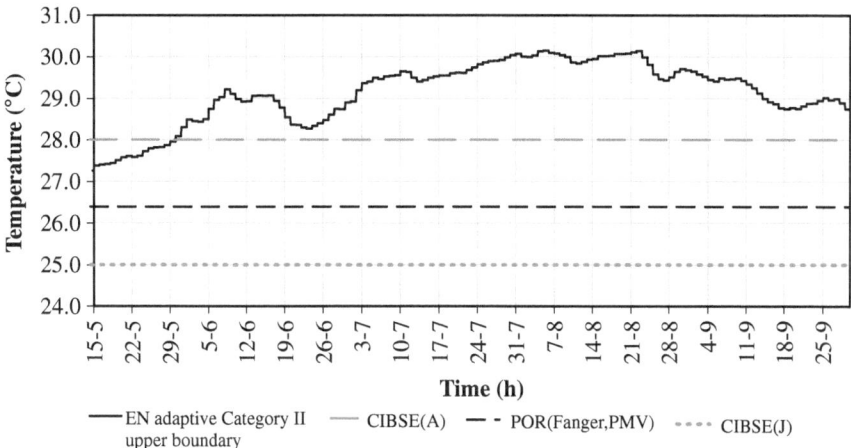

Fig. 2.12 Comparison of the upper boundary temperatures of POR and CIBSE indices calculated for the climate of Rome (IT)

modeled building is quite completely composed by offices similar by size, occupation rate and occupation schedule, Exceedance$_{PPD}$ is close to evaluate the frequency of exceedance of the PMV 0.85 threshold and Exceedance$_{Adaptive}$ is close to evaluate the frequency of exceedance of the ASHRAE adaptive comfort temperature rise of 3.5 °C. The main differences between the Exceedance indices and the Percentage outside range indices are that the former are asymmetric and are just used to assess summer overheating, while the latter are symmetric and can be used also for winter assessments. Their comparison follows in Fig. 2.10.

Considering the Category II and accounting only for the upper exceedance, POR$_{Fanger,PMV}$ considers the frequency of exceeding the threshold for PMV = +0.50, which is a stricter condition than Exceedance$_{PPD}$ threshold of PMV = 0.85; this explain why the values of POR$_{Fanger,PMV}$ are higher than Exceedance$_{PPD}$ (up to the double) and the apparent shift of the two trends.

POR$_{Adaptive}$ considers the frequency of exceedance of ±3.0 °C from EN adaptive comfort temperature, which is at first sight a stricter condition than the one used in Exceedance$_{Adaptive}$, where the exceedance threshold is set to +3.5 °C, but for most climates EN theoretical adaptive comfort temperature is higher than ASHRAE adaptive comfort temperature. When the difference between the two comfort temperatures is higher than 0.5 K, the ASHRAE adaptive threshold is lower than the EN adaptive threshold and this causes to Exceedance$_{Adaptive}$ to be higher than POR$_{Adaptive}$.

In summary, an interesting feature of Exceedance$_{PPD}$ and Exceedance$_{Adaptive}$ is the fact that they take into account occupancy. This implies that (i) they require a detailed description of the occupation schedules inside each zone of the building in order to be correctly applied, and (ii) if zoning is simplified they tend to coincide with measuring a percentage of hours outside comfort ranges, as it is the case in Percentage outside range indices. However, Exceedance$_{PMV}$ differs from

POR$_{Fanger,PMV}$ due to a different definition for the boundary values respectively of the acceptability class and of the comfort category; Exceedance$_{Adaptive}$ differs from POR$_{Adaptive}$ due to both the different definition for the boundary values respectively of the acceptability class and of the comfort category and the different optimal adaptive comfort temperatures which shall be calculated respectively according to ASHRAE 55 [50] and EN 15251.

2.3.2.4 CIBSE Criteria

The CIBSE criteria consider the time frequency by which dry-resultant temperature exceeds a specified reference temperature in summer: for CIBSE$_J$ this reference is 25 °C [46] whilst for CIBSE$_A$ it is 28 °C (26 °C for bedrooms) [45]. CIBSE$_A$ is suggested for naturally ventilated buildings. According to their definitions, CIBSE criteria are asymmetric.

Since CIBSE reports are proposed for being used in a worldwide network, we applied also CIBSE A and J criteria to the climate under investigation in this chapter; they are compared in Fig. 2.11 with the sole portions of POR$_{Fanger,PMV}$ and POR$_{Adaptive}$ representing overheating.

Both CIBSE criteria measure the frequency of overheating and do not take into account its severity, CIBSE$_J$ is relatively close to POR$_{Fanger,PMV}$, which is addressed to mechanically cooled buildings, and CIBSE$_A$ is relatively close to POR$_{Adaptive}$, which is addressed to not mechanically cooled buildings.

The upper PMV limit for Category II is set to 0.5 in EN 15251. Assuming that air temperature is approximately equal to mean radiant temperature, indoor air relative humidity of 50 %, air velocity of 0.1 m s^{-1}, metabolic activity of 1.2 met, clothing resistance of 0.5 clo and external work of zero met, a PMV = +0.5 (under such conditions) translates into an offset of + 1.7 °C from neutral temperature.

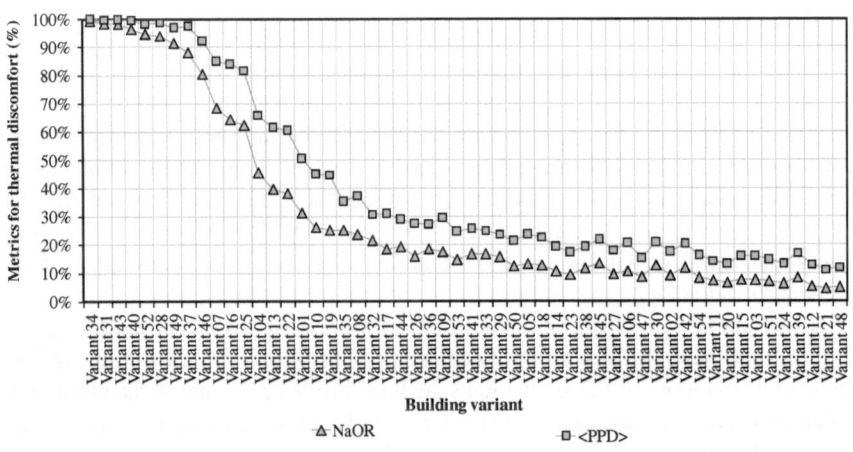

Fig. 2.13 Comparison of the indices directly derived from thermal sensation votes

Fig. 2.14 Correlation
between NaOR and <PPD>

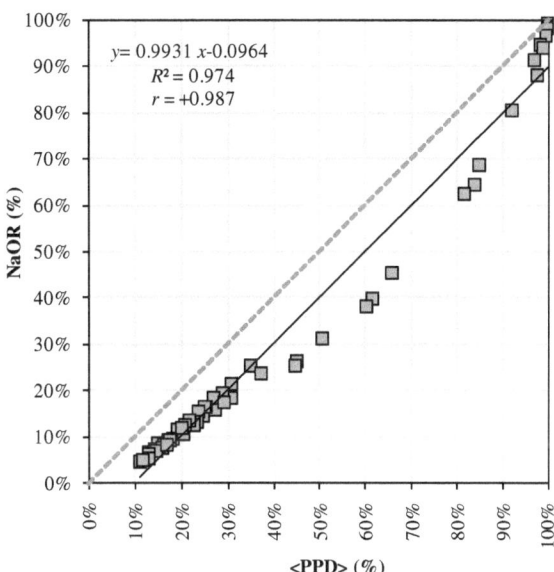

Hence the upper boundary operative temperature for $POR_{Fanger,PMV}$ is 26.4 °C. The CIBSE$_J$ criterion is hence generally stricter than Category II (Fanger) boundary expressed in operative temperature.

The $POR_{Adaptive}$ boundary depends on the exponential weighted running-mean dry bulb temperature of outdoor air [2], and hence it depends on climate. Fig. 2.12 presents a comparison of boundary temperatures for the climate of Rome (IT).

Obviously the shift of the reference temperature from 25 to 28 °C causes a strong change in trend and in the values of the CIBSE indices. Some authors, based on a large database of measurements and interviews, have found evidence that, in naturally ventilated buildings or in buildings without a mechanical cooling system, summer comfort temperature varies with the outdoor dry-bulb air temperature [51, 52], and this result has been translated into parts of the ASHRAE and EN standards. Thus, indices based on a fixed reference temperature seem to be a somewhat crude approximation for assessing free-running buildings (naturally ventilated or without a mechanical cooling system) and are also a simplification for assessing mechanically cooled building.

It should be noted that Fanger summer comfort temperature does not depend explicitly on external climate, but it may vary considerably depending on choices made by building occupants about insulation levels provided by clothing and chairs, air velocity, etc., which in turn may be correlated to external climate. In the words of Fanger and Toftum:

The PMV model has been referred to as a static model, indicating that it should prescribe one constant temperature. But this is not true. The PMV model may actually predict any air temperature between 10 and 35 °C as neutral, depending on the other five variables in the model [53].

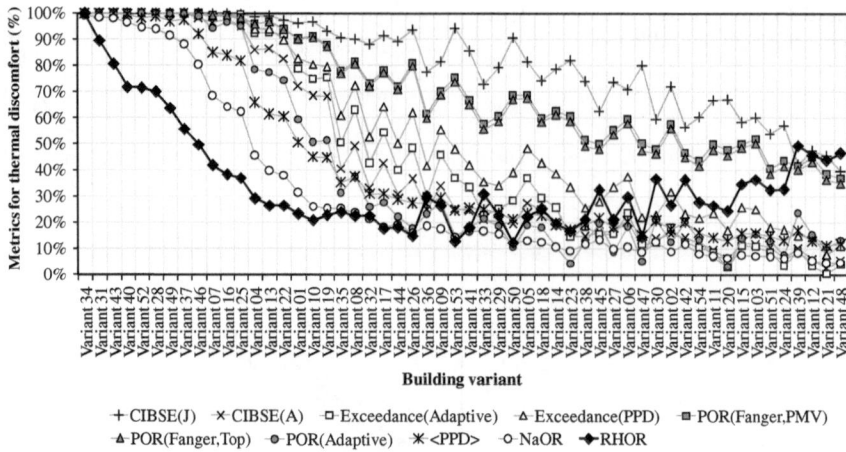

Fig. 2.15 Comparison of the indices that can be represented on the percentage scale

Thus, indices based on a fixed reference temperature are not reliable for assessing free-running buildings (naturally ventilated or without a mechanical cooling system) and are a simplification for assessing mechanically cooled building. Often reference temperatures could be directly calculated from the Fanger comfort model.

2.3.3 Indices that Explicitly Make Use of Likelihood of Dissatisfied

The relationship between human thermal discomfort indices and indoor temperature depends on thermal sensation of occupants and it is not linear. In Fanger's words, *human thermal discomfort* can be translated with *predicted percentage of dissatisfied* (PPD) whilst *thermal sensation* with *predicted mean vote* (PMV). The relationship between PPD and PMV is an exponential function [54]. de Dear and Brager [52, 55] used the same relationship in order to correlate *thermal sensation votes* and *percentage of dissatisfied* in their analysis of a large worldwide database of thermal comfort surveys. Nicol and Humphreys [56] showed by analysis of comfort databases that *offset from comfort temperature* is related to the *percentage preference* (the complementary of human thermal discomfort expressed as a percentage) through a logistic function.

PPD was introduced by Fanger [53] and derived by the analysis of the thermal sensation votes of people placed in thermostatic chambers. The NaOR index was derived from a logistic regression analysis [7] of the data collected in European office buildings during the SCATs Project [47].

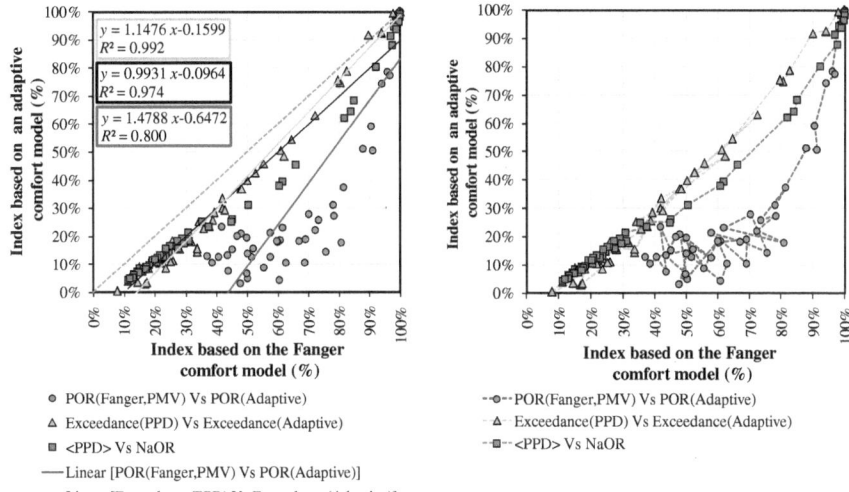

Fig. 2.16 *Left* Linear regression analyses and *Right* optimization paths of indices based on Fanger and adaptive comfort models

Both NaOR and <PPD> are short-term discomfort indices, and in order to use them for a long-term evaluation, they have been averaged over the occupied hours in the specified calculation period. Thus, the average values have been calculated for the 54 building variants in order to compare them with other indices. Figure 2.13 reports the comparison between the long-term NaOR and <PPD>.

Compared to all other indices expressed on a percentage scale, both NaOR and <PPD> fluctuate less than the other indices when the variants are ordered with respect to the average indoor operative temperature.

Even if their values are different, their trends are remarkably similar, and they select the same variant as optimal. To evaluate the strength of their relationship, a Pearson correlation analysis was performed assuming a linear correlation between them. The correlation is strong (Pearson correlation coefficient, $r > +0.98$), and the linear trend fits the data with a good agreement ($R^2 > 0.97$). At least in the presented case study, the linear correlation is particularly high for building variants characterized by low discomfort (minor than 30 %), which are the ones more relevant for optimization processes towards net zero energy buildings (Fig. 2.14).

Summarizing, the strengths of <PPD> and NaOR are: (i) they are based on two underlying accepted comfort models and so they explicitly take into account the predicted discomfort sensation of occupants; (ii) they assess building performance in a similar way and so they provide similar design suggestions and would not create relevant discontinuities when applying the two-step optimization process proposed in EN 15251; and (iii) they can be simply calculated from hourly simulated and measured data.

On the other side, in the version here used, their weaknesses are: (i) they do not account for occupation rate and (ii) <PPD> is symmetrical whilst NaOR is asymmetrical; thus they response differently in those situations when overcooling may be predominant, so their complementarity breaks down in those situations.

2.3.3.1 Robinson's and Haldi's Overheating Risk

RHOR was developed for predicting the hourly likelihood of summer overheating and we have evaluated its seasonal average over the sole occupied hours, in order to compare it with other indices. Such index requires to set the values of two time constants a and b: a was set to $4.75 \cdot 10^{-4}$ and b was set to $4.75 \cdot 10^{-3}$ as suggested in [8].

The overall comparison of the percentage indices shows that RHOR, when compared with all other indices, provides comparatively lower discomfort values for those variants with higher indoor operative temperatures and tends to comparatively higher discomfort values in those variants with lower indoor operative temperatures (Fig. 2.15).

As suggested by its authors, this index needs to tune both the time constants for charging and for discharging in relation with the specific building features and the local climate. We hypnotize that the used values for the two time constants might be the cause of its trend considerably different by the other indices. This hypothesis, whether confirmed, would imply that RHOR is difficult to generalize.

2.3.4 Inter-Comfort Model Correlation

In order to identify the pairs of indices based on the two comfort models that are more correlated and more suitable for complementary use, we propose to use the degree of linear correlation between indices across the comfort models. The indices selected for the inter-comfort-model analysis are those expressed in percentage and applicable to both Fanger and adaptive comfort models. The pairs that have been identified are:

1. $POR_{Fanger,PMV}$ and $POR_{Adaptive}$.
2. $Exceedance_{PPD}$ and $Exceedance_{Adaptive}$.
3. <PPD> and NaOR.

The pairs of indices are evaluated over the same 54 variants of the aforementioned office building (Fig. 2.16).

The inter-comfort-model-correlation analysis shows that:

- Pair #1: the assumption of linearity between $POR_{Fanger,PMV}$ and $POR_{Adaptive}$ is weak ($R^2 = 0.80$), and the two evaluation procedures are significantly different; $POR_{Adaptive}$ does not have a monotone behavior in response to changes of $POR_{Fanger,PMV}$ (see right side of Fig. 2.16).

- Pair #2 indicates that its indices similarly evaluate the relative performance of different variants (gradient = 0.99) and is well represented by a linear relationship ($R^2 = 0.97$) in particular for discomfort values smaller than 30 % ($R^2 = 1.00$); NaOR generally has a monotone behavior in response to changes of <PPD>.
- Pair #3 is the best fitted by a linear regression ($R^2 = 0.99$), but Exceedance$_{PPD}$ rises 15 % more steeply than Exceedance$_{Adaptive}$, and Exceedance$_{Adaptive}$ does not have a monotone behavior in response to changes of Exceedance$_{PPD}$.

The previous analysis brings us to the conclusion that <PPD> and NaOR might be the most suitable indices for having correlated complementary thermal comfort assessments based on the Fanger and the EU adaptive comfort models. They are potential candidates to be used in the two-step optimization procedure suggested by EN 15251. Since they provide a similar ranking of variants, their complementary use might avoid the creation of the unnecessary discontinuities, which are currently present in the optimization process proposed in EN 15251, when switching from an adaptive-based index to a Fanger-based one [3]. On the other hand, one should notice that <PPD> and NaOR account differently for those situations when overcooling may be predominant, so their complementarity breaks down in those cases.

2.4 Conclusions

In this chapter, 16 long-term discomfort indices have been used for assessing the long-term thermal comfort performances of 54 alternative variants of a 34-zone office building during the summer season. The objective was to assess the ranking capability of such indices in order to be reliably used in two-step optimization procedure proposed in EN 15251 (first optimize envelope and passive systems towards an adaptive-based long-term discomfort index, and whether comfort cannot be met, then optimize envelope and active systems towards a Fanger-based long-term discomfort index). The tested indices are not fully reliable, and the most of them produce a discontinuity when switching the comfort model.

A number of remarks and suggestions have been drawn from this analysis and here we summarize the ones most relevant for practical use and for further developments:

- In many instances, the indices provide significantly different conclusions about thermal comfort performances and ranking of building variants. The indices show trends over the 54 variants (ordered by decreasing space and time averaged operative temperature) that are different in absolute value and shape. They would deliver different guidance conclusions to a designer for the optimization of a building.
- Based on their features and mutual correlation, <PPD> and NaOR might be, among the existing indices, the most suitable couple of indices for having

correlated complementary thermal comfort assessments based on the Fanger and the European adaptive comfort models. They are promising candidates to be used in the two-step optimization procedure suggested by EN 15251 since they provide a similar ranking of variants, and might avoid the creation of a discontinuity when switching from an adaptive-based index to a Fanger-based index. On the other hand, one should notice that <PPD> and NaOR account differently for those situations when overcooling may be predominant, so their complementarity breaks down in those cases.

- Symmetric indices may be useful to represent the overall discomfort performance of a building in a comprehensive way since they account both for overheating and overcooling phenomena. However, their disaggregation in overheating and overcooling should be explicitly reported (e.g., in the format of Fig. 2.9), in order to allow for a complete and correct interpretation. Further studies based on comfort surveys should address the question whether summer overheating and overcooling are, in fact, perceived with the same *discomfort weight* by building occupants, also in connection to personal control possibilities (e.g., clothing adjustment, air velocity adjustment, etc.).

- Indices based on a fixed reference temperature seem to be a somewhat crude approximation for assessing free-running buildings (naturally ventilated or anyway without a mechanical cooling system) where on the contrary analysis of available comfort databases points to dependence of summer comfort temperature from external climate [55]. They are an approximation also when assessing mechanically cooled buildings since the Fanger summer comfort temperature is not explicitly dependent on external climate, and may vary considerably depending on choices made by building occupants about insulation levels provided by clothing and chairs, air velocity, etc. [54], which in turn may be correlated to external climate.

- Since $POR_{Fanger,PMV}$ is a simple counter of the hours outside the comfort range, while <PPD> provides a measure of human thermal perception by accounting for severity of discomfort in each hour, we suggest that <PPD> should be preferred, because it is more accurate than $POR_{Fanger,PMV}$ to assess thermal discomfort according to the Fanger comfort model.

- ISO 7730 and EN 15251 propose either the PMV or operative temperature as options for the calculation of the Percentage outside range indices, based on the Fanger comfort model. The two indices show little differences since simulated hourly PMV also depends on hourly relative humidity, which is assumed constant when traducing PMV in operative temperature, but humidity has only a little influence on thermal comfort temperature in moderate environments according to ISO 7730.

- Exceedance$_{PPD}$ and Exceedance$_{Adaptive}$, by taking into account the occupation rate for each zone and time interval, give a more accurate description of overall discomfort with respect to those indices that weigh on floor area or volume, irrespective of occupation. Thus, the Exceedance indices require a detailed zoning of the building and a detailed description of the occupation schedules inside each zone in order to be used. In case the hourly occupation rate is almost constant in each zone of the building or if zoning is extremely simplified, Exceedance$_{PPD}$ and Exceedance$_{Adaptive}$ converge to measure the frequency of exceedance of the specified comfort boundary; however, Exceedance$_{PMV}$ differs from POR$_{Fanger,PMV}$ due to a different definition for the boundary values respectively of the acceptability class and of the comfort category; Exceedance$_{Adaptive}$ differs from POR$_{Adaptive}$ due to both the different definition for the boundary values respectively of the acceptability class and of the comfort category and the different optimal adaptive comfort temperatures, which shall be calculated respectively according to ASHRAE 55 and EN 15251.

References

1. ISO (2005) ISO 7730: Ergonomics of the thermal environment—analytical determination and interpretation of thermal comfort using calculation of the PMV and PPD indices and local thermal comfort criteria. 3rd version. International Standard Organization (ISO), Geneva
2. CEN (2007) EN 15251: Indoor environmental input parameters for design and assessment of energy performance of buildings addressing indoor air quality, thermal environment, lighting and acoustics. European Committee for Standardization (CEN), Brussels
3. L. Pagliano, P. Zangheri, Comfort models and cooling of buildings in the Mediterranean zone. Adv. Build. Energy Res. **4**, 167–200 (2010)
4. W. Pane, Thermal mass and overheating—provide modeling support to guidance on overheating in dwellings. BRE (2005), http://www.sipcrete.com/PDF%20Downloads/Thermal_Mass_and_Overheating.pdf. Accessed 9 Feb 2013
5. J. Schnieders, *Passive Houses in South West Europe* (Passivhaus Institut, Darmstadt, 2009)
6. W. Feist, R. Pfluger, B. Kaufmann, J. Schnieders, O. Kah, *Passive House Planning Package 2004.* (Passive House Institute, Darmstadt, 2004)
7. J.F. Nicol, J. Hacker, B. Spires, H. Davies, Suggestion for new approach to overheating diagnostic, in *Proceedings of Conference: Air Conditioning and the Low Carbon Cooling Challenge*, Cumberland Lodge, Windsor, 2008
8. D. Robinson, F. Haldi, Model to predict overheating risk based on an electrical capacitor analogy. Energy Build. **40**, 1240–1245 (2008)
9. L. Grignon-Massé, D. Marchio, M. Pietrobon, L. Pagliano, Evaluation of energy savings related to building envelope retrofit techniques and ventilation strategies for low energy cooling in offices and commercial sector, in *Proceedings of the 6th International Conference on Improving Energy Efficiency in Commercial Building*, Frankfurt 2010
10. S. Borgeson, G.S. Brager, Comfort exceedance metrics in mixed-mode buildings, in *Proceedings of Conference: Adapting to Change: New Thinking on Comfort Cumberland Lodge*, Windsor (2010)
11. C. Di Perna, F. Stazi, A. Ursini Casalena, M. D'Orazio, Influence of the internal inertia of the building envelope on summertime comfort in buildings with high internal heat loads. Energy Build. **43**, 200–206 (2011)
12. M.C. Mozer, L. Vidmar, R.H. Dodier, The neurothermostat: predictive optimal control of residential heating systems, in *Proceedings of Advances in Neural Information Processing Systems* 9, (MIT Press, Cambridge, 1997) pp. 953–959
13. S. Wang, X. Jin, Model-based optimal control of VAV air-conditioning system using genetic algorithm. Build. Environ. **35**, 471–487 (2000)
14. A. Angelotti, L. Pagliano, G. Solaini, Summer cooling by earth-to-water heat exchangers: experimental results and optimisation by dynamic simulation, in *Proceedings of EuroSun 2004*, Freiburg (2004) pp. 678–686

15. C. Park, G. Augenbroe, N. Sadegh, M. Thitisawat, T. Messadi, Real-time optimization of a double-skin façade based on lumped modeling and occupant preference. Build. Environ. **39**, 939–948 (2004)
16. N. Nassif, S. Kajl, R. Sabourin, Two-objective on-line optimization of supervisory control strategy. Build. Serv. Eng. **25**(3), 241–251 (2004)
17. N. Nassif, K. Stainslwa, R. Sabourin, Optimization of HVAC control system strategy using two-objective genetic algorithm. HVAC&R Res. **11**(3), 459–486 (2005)
18. M. Kummert, P. André, Simulation of a model-based optimal controller for heating systems under realistic hypothesis, in *Proceedings of the 9th IBPSA Conference—Building Simulation*, Montreal (2005)
19. M. Emmerich, C. Hopfe, R. Marijt, J. Hensen, C. Struck, L. Stoelinga, Evaluating optimization methodologies for future integration in building performance tools, in *Proceedings of the 8th International Conference on Adaptive Computing in Design and Manufacture*, Bristol (2008) pp. 1–7
20. M. Mossolly, K. Ghali, N. Ghaddar, Optimal control strategy for a multi-zone air conditioning system using a genetic algorithm. Energy **34**, 58–66 (2009)
21. C.D. Corbin, G.P. Henze, P. May-Ostendorp, A model predictive control optimization environment for real-time commercial building application. J. Build. Perform. Simul. (2012)
22. R. Loonen, M. Trcka, J. Hensen, Exploring the potential of climate adaptive building shells, in *Proceedings of the 12th Conference of International Building Performance Simulation Association*, Sydney (2011) pp. 2148–2155
23. L. Stephan, A. Bastide, E. Wurtz, Optimizing opening dimensions for naturally ventilated buildings. Appl. Energy **88**, 2791–2801 (2011)
24. D. Kolokotsa, G.S. Stavrakakis, K. Kalaitzakis, D. Agoris, Genetic algorithms optimized fuzzy controller for the indoor environmental management in buildings implemented using PLC and local operating networks. Eng. Appl. Artif. Intel. **15**, 417–428 (2002)
25. L. Magnier, F. Haghighat, Multiobjective optimization of building design using TRNSYS simulations, genetic algorithms, and artificial neural network. Build. Environ. **45**, 739–746 (2010)
26. P. Hoes, J.L.M. Hensen, M.G.L.C. Loomans, B. Vries, D. de Bourgeois, User behavior in whole building simulation. Energy Build. **41**, 295–302 (2009)
27. M. Hamdy, A. Hasan, K. Siren, Applying a multi-objective optimization approach for design of low-emission cost-effective dwellings. Build. Environ. **46**, 109–123 (2011)
28. P. Hoes, M. Trcka, J.L.M. Hensen, B. Hoekstra Bonnema, Optimizing building designs using a robustness indicator with respect to user behavior, in *Proceedings of the 12th Conference of the International Building Performance Simulation Association*, Sydney (2011) pp. 1710–1717
29. D.B. Crawley, L.K. Lawrie, C.O. Pedersen, F.C. Winkelmann, M.J. Witte, R.K. Strand, R.J. Liesen, W.F. Buhl, Y.J. Huang, R.H. Henninger, J. Glazer, D.E. Fisher, D.B. Shirey III, B.T. Griffith, P.G. Ellis, L. Gu, Energyplus: new, capable, and linked. J. Archit. Plan. Res. **21**(4), 1–10 (2004)
30. GUR, Decreto legislativo n. 311. Disposizioni correttive ed integrative al decreto legislativo 19 agosto 2005, n. 192, recante attuazione della direttiva 2002/91/CE, relativa al rendimento energetico nell'edilizia. Gazzetta ufficiale della Repubblica Italiana (2006)
31. SIA, SIA 180—Isolamento termico e protezione contro l'umidità degli edifice. Swiss Association of Architects and Engineers, Zurich (2009)
32. CEN, EN 13786, Thermal performance of building components—Dynamic thermal characteristics—Calculation methods, European Committee for Standardization, Brussels (2007)
33. ASHRAE: ANSI/ASHRAE Standard 140—Standard Method of Test for the Evaluation of Building Energy Analysis Computer Programs. American Society of Heating, Refrigerating and Air-Conditioning Engineers, Atlanta (2011)

34. G. Beccali, M. Cellura, M. Lo Brano, A. Orioli, Is the transfer function method reliable in a EN building context? A theoretical analysis and a case study in the south of Italy. Appl. Therm. Eng. **25**, 341–357 (2005)
35. US DOE, EnergyPlus Engineering Reference: The Reference to EnergyPlus Calculations. U.S. Department of Energy (2011)
36. S. Fohanno, G. Polidori, Modelling of natural convective heat transfer at an internal surface. Energy Build. **38**, 548–553 (2006)
37. F. Alamdari, G.P. Hammond, Improved data correlations for buoyancy-driven convection in rooms. Build. Serv. Eng. Res. Technol. **4**(3), 106–112 (1983)
38. G.N. Walton, *Thermal Analysis Research Program Reference Manual*. NBSSIR 83- 2655 (National Bureau of Standards, Washington, 1983)
39. ISO, ISO 15099—Thermal performance of windows, doors, and shading devices—detailed calculations. International Standard Organization (2003)
40. B. Givoni, Options and applications of passive cooling. Energy Build. **7**, 297–300 (1984)
41. G.N. Walton, *AIRNET—A Computer Program for Building Airflow Network Modeling. NISTIR 89-4072* (National Institute of Standards and Technology, Gaithersburg, 1989)
42. W.S. Dols, G.N. Walton, *CONTAM 2.0 User Manual. NISTIR 6921* (National Institute of Standards and Technology, Gaithersburg, 2002)
43. G.N. Walton, W.S. Dols, *CONTAM 2.1 Supplemental User Guide and Program Documentation. NISTIR 7049* (National Institute of Standards and Technology, Gaithersburg, 2003)
44. A. Mitra, *Fundamentals of Quality Control and Improvement*, 2nd edn. (Prentice Hall, New Jersey, 1998)
45. CIBSE, *Guide A—Environmental Design* (Chartered Institution of Building Services Engineers, London, 2006)
46. CIBSE, *Guide J—Weather, solar and illuminance data* (Chartered Institution of Building Services Engineers, London, 2002)
47. J.F. Nicol, K. McCartney, *Final report of Smart Controls and Thermal Comfort (SCATs) Project. Report to the European Commission of the Smart Controls and Thermal Comfort project* (Oxford Brookes University, UK, 2001)
48. A. Buda, A. Jarynowski, Life-time of correlations and its applications. Wydawnictwo Niezalezne **1**, 5–21 (2010)
49. J. Cohen, *Statistical Power Analysis for the Behavioral Sciences*, 2nd edn. (Department of Psychology, New York University, Lawrence Erlbaum Associates, Hillsdale, 1998)
50. ASHRAE, *ANSI/ASHRAE 55—Thermal Environmental Conditions for Human Occupancy* (American Society of Heating, Refrigerating and Air-Conditioning Engineers, Atlanta, 2004)
51. J.F. Nicol, M.A. Humphreys, Adaptive thermal comfort and sustainable thermal standards for buildings. Energy Build. **34**, 563–572 (2002)
52. R.J. de Dear, G.S. Brager, D. Cooper, Developing an Adaptive Model of Thermal Comfort and Preference—Final report ASHRAE RP-884 (1997)
53. P.O. Fanger, J. Toftum, Extension of the PMV model to non-air-conditioned building in warm climates. Energy Build. **34**, 533–536 (2002)
54. P.O. Fanger, *Thermal comfort* (Danish Technical Press, Copenhagen, 1970)
55. R.J. de Dear, G.S. Brager, Towards an adaptive model of thermal comfort and preference. ASHRAE Trans. **104**(1), 145–167 (1998)
56. J.F. Nicol, M.A. Humphreys, Maximum temperatures in EN office buildings to avoid heat discomfort. Sol. Energy **81**, 295–304 (2007)
57. SIA, *SIA 380/1—L'energia termica nell'edilizia* (Swiss Association of Architects and Engineers, Zurich, 2009)

Chapter 3
Gap Analysis of the Long-Term Discomfort Indices and a Harmonized Calculation Framework

Abstract Long-term discomfort indices depend on a number of boundary parameters that are currently not univocally defined in standards, although they have a large influence on the results. Specifically, EN 15251 and ASHRAE 55 do not clarify some gaps and provide a wide margin of discretion to the analyst when calculating them. In order to assess gaps and needs for using long-term discomfort indices, a selection of such indices has been tested, during a summer period, by means of a sensitivity analysis. It was found that the sensitively to given gaps is higher for those building variants characterized by lower discomfort levels. In order to make the long-term discomfort indices reliable tools (i) for establishing requirements about the indoor thermal environment of buildings, (ii) for driving the comfort-optimization of buildings and (iii) for assessing their comfort performance during the operational phase, the identified gaps shall be fixed, and it is desirable that they are framed in a common and accepted standard framework. Finally, a calculation framework is proposed, and a new method for calculating the extension of the calculation period (seasons) is also presented.

3.1 Introduction

The use of indices for assessing the long-term thermal comfort performance of a building is growing among the scientific community. Such indices have been proposed by standards and pieces of literature over the last 10 years and are useful tools that synthesize the long-term thermal comfort performance of a building in a single value. They express requirements about the indoor building environment, for comparing different passive or active design strategies and thus supporting the selection of the most suitable ones. However, such indices depend on a number of boundary parameters that are currently not univocally defined in Standards;

thus they let the analyst a wide margin of discretion when calculating long-term discomfort indices. A literature review allows to identify the main gaps influencing their calculation.

3.2 Literature Review

Long-term discomfort indices are mathematical formulations that integrate the exceedance from an assumed thermal comfort condition over a given period. Standards and guidelines (quite) completely define the procedures for calculating such comfort conditions, commonly expressed as a range around a comfort temperature or neutrality. However, some aspects of the procedures for calculating the indices are not fully stated.

The three main gaps are (i) the definition of the calculation period, (ii) the boundary temperatures of comfort categories for the Fanger model as proposed in EN 15251 [1] and (iii) the input meteorological variable for calculating the comfort temperature according to ASHRAE 55 [2].

3.2.1 Uncertainty About the Calculation Period

Since most of these indices are calculated by averaging or by accumulating over time, one of the most influential boundary condition is the calculation period, specifically the duration and beginning of summer and winter. The standards ISO 7730 [3] and EN 15251 propose respectively five and three methods to assess long-term uncomfortable conditions and refer explicitly to summer and winter or warm and cold periods, but they do not provide a standardized method to univocally identify the starting day and the duration of the seasons. Also, the standard ASHRAE 55 does not specify how to explicitly identify summer and winter periods. CIBSE introduced a pair of criteria for avoiding summer overheating in the Guide J [4] and Guide A [5], but a method for identifying summer is not available. However, arbitrarily increasing the duration of the calculation period may reduce the values of the long-term discomfort indices [6] since a higher number of hours when outdoor climatic stresses are weaker are included.

Different approaches to define seasons can be found in the scientific literature: (i) direct definitions, (ii) indirect definitions based on outdoor air temperature, (iii) indirect definition based on the interpretation of people's clothing choice.

A number of direct definitions of the seasons are available. According to the *meteorological definition* [7], the seasons are periods having three calendar months. The *astronomical definition* [8] defines winter as the period from the winter solstice to the vernal equinox, spring ends at the summer solstice; summer lasts from the summer solstice until the autumnal equinox; autumn from the autumnal equinox until the winter solstice. Due to the elliptical orbit of the Earth around the Sun, the duration of astronomical seasons varies from 89 to 93 days. Both these definitions of seasons are not able to represent local climatic conditions [9]. Climatologists have proposed other definitions, e.g., *temperature definition* [9], *synoptic definition* [8]. Such definitions are closer to synthesize climatic peculiarities of a local climate, but both require quite complex calculations and are not suitable to be related with thermal comfort conditions in buildings. de Dear and Brager in a first proposal for an adaptive comfort standard state that *summer* could be:

> [...] operationally defined as the cooling season; climatologically defined for the purposes of this standard as having a mean daily outdoor effective temperature of 25 °C [10].

and *winter* could be:

> [...] operationally defined as the heating season; climatologically, for the purposes of this standard, a typical winter condition is assumed to have a mean daily outdoor effective temperature of 0 °C [10].

However, we think that these two definitions are not reliable since a season cannot be represented by a single value, but it should be defined through an interval of temperatures. Moreover, since adaptive models are allowed to be used in free-running buildings, the expression *cooling season* is semantically not suitable because it assumes the presence of a mechanical cooling system for controlling the indoor environment.

Indirect definitions derive from comparing a function of outdoor air temperature with the comfort ranges proposed by the adaptive comfort models.

The ASHRAE adaptive model is provided for the range of mean monthly outdoor air temperature between 10 and 33 °C (Fig. 3.1); hence, the warm period (or summer) might implicitly correspond to the period of the year when the outdoor climate falls into this range, and the cold period (or winter) might be assumed to be the period that completes the annual cycle.

Fig. 3.1 Acceptable temperature ranges for naturally conditioned spaces (Data from [2])

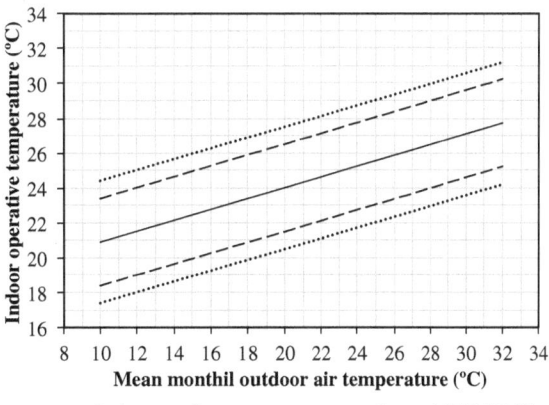

For the EN Adaptive model, the situation is more complex: it defines upper comfort boundaries for a running mean indoor air temperature from 10 to 30 °C and lower comfort boundaries from 15 to 30 °C (Fig. 3.2).

In order to understand this difference in the temperature boundaries, we refer to the category ranges proposed by Olesen et al. [11] (Fig. 3.3). Although they use a different meteorological metric as independent variable, the upper and lower boundaries of the comfort ranges are different in order to match the *summer trend* (correlated to the outdoor temperature) with the *winter trend* (independent of the outdoor temperature). According to this interpretation, *warm period* (or *summer*) may begin when the lower boundary start to be correlated to the outdoor air temperature, i.e., for a mean monthly outdoor air temperature higher than 18 °C, winter when it is lower than 10 °C and intermediate seasons between 10 ÷ 18 °C.

Applying this interpretation to the EN 15251 graph, the running mean of the outdoor air temperature of 15 °C can be considered the *switching temperature* between summer and winter. Such three approaches are valid until the independent variable falls into the specified temperature ranges. When the independent variable is outside those ranges, there are no suggestions on how to proceed and, therefore, the aforementioned definitions of *warm* and *cold periods* formally decade. This is obviously a limit of these indirect definitions. Another limit consists in leading, in some cases, to a noncontiguous set of days, i.e., the independent variable may fluctuate around the *switching temperature* in consecutive days.

The third attempt to define seasons consists in interpreting people's clothing choice. Haldi and Robinson [12] showed that insulation level of clothing is correlated with a good agreement ($R^2 = 0.97$ obtained on binned data) to the running mean of outdoor air temperature and that people choose clothing more in a predictive way, based on the experienced evolution of outdoor dry-bulb air temperature, than like a consequence of the actual indoor thermal environment. According to Haldi and Robinson, a clothing resistance of 0.5 clo (used in

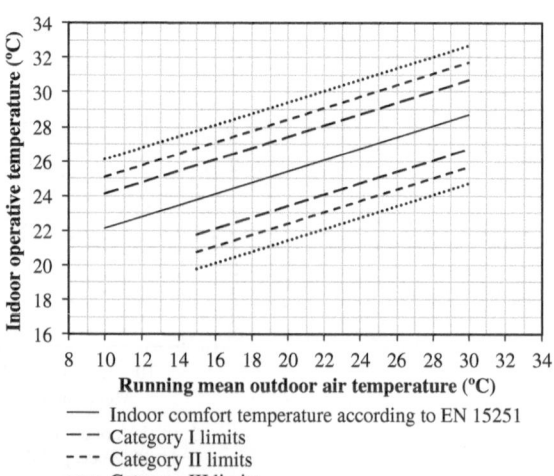

Fig. 3.2 Design values for the indoor operative temperature for buildings without mechanical cooling systems as a function of the running mean outdoor air temperature (Data from [1])

Fig. 3.3 Design values for
the indoor operative
temperature as a function of
mean monthly outdoor air
temperature for buildings
without mechanical cooling
systems (Data from Olesen
et al. [11])

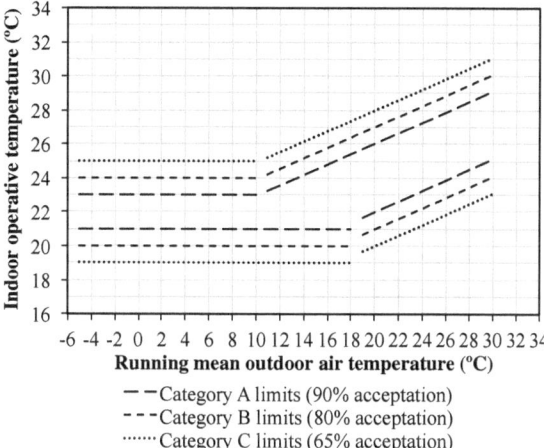

standards for offices during the summer) is correlated to a running mean of outdoor
air temperature of 22 °C, while a clothing resistance of 1.0 clo (used in standards
for offices during winter) is correlated to a running mean of outdoor air temperature of 10 °C; thus *summer* may be defined as the period when running mean
temperature is higher than 22 °C and *winter* when the running mean temperature is
lower than 10 °C and the *intermediate seasons* when the running mean temperature is in the interval 10 ÷ 22 °C. Kalz and Pfafferott [13], analyzing the results
of studies on people's clothing behavior, formulated the hypothesis that a
switching temperature from winter to summer is a running mean of outdoor air
temperature of 15 °C for which the clothing level is around 0.7 clo. However, also
these two approaches suffer for the limit to isolate noncontiguous set of days with
the running mean temperature stable above or below the proposed thresholds and
the daily fluctuations and anomalous climatic phenomena might disturb the clear
identification of the starting and ending days of the calculation period.

3.2.2 Rounded Boundary Temperatures of Comfort Categories in EN 15251

According to ISO 7730 and EN 15251, the boundaries of the comfort categories
based on Fanger model are defined in terms of PMV, but they are also translated
into operative temperature, by making certain assumptions about the values of
external work rate, metabolic rate, clothing resistance, air velocity, relative
humidity and the relationship between dry-bulb air temperature, mean radiant
temperature and operative temperature. Although EN 15251 refers to ISO 7730 for
the Fanger model, and the PMV limits of the categories are the same for both these
standards (for Category I, $PMV_{limit} = \pm 0.2$; for Category II, $PMV_{limit} = \pm 0.5$;
for Category III, $PMV_{limit} = \pm 0.7$), the boundary operative temperatures proposed

in the table A.3 of EN 15251 are different from those reported in the table E.3 of ISO 7730, even if they are subject to similar assumptions. Therefore, referring to ISO 7730 or EN 15251 implies a difference in the boundary operative temperatures of comfort categories that causes a difference in the values of those long-term discomfort indices that depend on comfort categories (see Sect. 2.3.2).

3.2.3 Uncertainty About the Meteorological Input Variable in ASHRAE 55

The American standard ASHRAE 55 defines the mean monthly outdoor dry-bulb air temperature as:

> [...] when used as input variable in Fig. 5.3.1 for the adaptive model, this temperature is based on the arithmetic average of the mean daily minimum and mean daily maximum outdoor (dry-bulb) temperatures for the month in question [2].

Borgeson and Brager refer, instead, to a "mean outdoor dry bulb temperature for the previous month" [14] and Huizenga states: "ASHRAE Standard 55-2004 uses a 30 day averaging period" [15]. Thus, it is not clear if to take as reference the current calendar month, the previous month or the last running 30 days for the averaging operation. This uncertainty is reinforced by the most recent release of the building performance simulation software EnergyPlus, the version 7, which includes new thermal comfort outputs based on the adaptive comfort models based on both ASHRAE 55 and EN 15251. In fact, according to such release, the meteorological input for the ASHRAE adaptive model can be supplied to the software or through a .stat file containing the daily average temperatures for each calendar month or, if no .stat file is provided, the software calculates the monthly mean outdoor temperature as a running average of the previous 30-day average temperatures taken from the weather file [16]. The weak definition of the meteorological input causes two different curves for the optimal comfort temperature and, consequently, the values of those long-term discomfort indices depending on the adaptive ASHRAE optimal comfort temperature change.

3.3 Gap Analysis of a Selection of Long-Term Discomfort Indices

At first, a selection of the most representative long-term discomfort indices has been collected among those reviewed in Chap. 1. Only the indices based on the comfort models have been selected since they are able to assess the predicted thermal comfort perceived by a group of people. In particular, in the test about the modification of the calculation period, Percentage outside PMV range ($POR_{Fanger,PMV}$) and Percentage outside the adaptive temperature range ($POR_{Adaptive}$) have been selected since they measure the frequency of occurrence of a discomfort phenomenon respectively according to the Fanger and the Adaptive comfort models. The Percentage outside Fanger temperature range ($POR_{Fanger,op}$) has been excluded since it is very close to $POR_{Fanger,PMV}$ (see Sect. 2.3.2); Average PPD (<PPD>) has been selected since it measures the average predicted percentage of people that would feel uncomfortable according to the Fanger model and the Nicol et al.'s Overheating risk (NaOR) has been selected since it measure the average predicted percentage of people that would feel uncomfortable according to the EN adaptive model. Indices that cannot be represented by a percentage are not reported since PPD-weighted criteria (PPDwC) are similar to Accumulated PPD (Sum_PPD) and the latter is strictly linked to <PPD> (see Sect. 2.3.1). In the test about the difference in the Fanger boundary temperatures, Percentage outside Fanger temperature range ($POR_{Fanger,\theta op}$) and the Degree-hour criteria (DhC) have been selected. In the test about the meteorological input for the ASHRAE adaptive comfort model, the $Exceedance_{Adaptive}$ has been only used.

In order to assess the sensitively of the long-term discomfort indices to modifications of the identified gaps, the relative error, E_{Rel}, of the modified case with respect to the reference case is used as indicator $E_{Rel} = (V_{Test} - V_{Ref})/V_{Ref}$, where V indicates the value of a certain long-term discomfort index.

3.3.1 Modification of the Calculation Period

As aforementioned, most of the long-term discomfort indices are calculated by averaging or by accumulating values over a specified calculation period. Hence, a change in its duration or its shift result in a change of the value of the indices.

In order to evaluate the sensitively of a selection of representative long-term discomfort indices, (i) the duration of the calculation period, and (ii) the shift of the calculation period according to the scenarios proposed in Table 3.1 and (iii) the modification of the daily number of occupation hours of the building are changed.

Table 3.1 Graphical representation of the tested calculation periods

ID	May		June		July		August		September		October	
	1 - 14	15 - 31	1 - 14	15 - 30	1 - 14	15 - 30	1 - 14	15 - 30	1 - 14	15 - 30	1 - 14	15 - 30
	Duration of the calculation period											
Test A.1												
Reference period												
Test A.2												
	Displacement of the calculation period											
Test B.1												
Reference period												
Test B.2												

3.3.1.1 Duration of the Calculation Period

As suggested by Nicol et al. [6], an overheating problem detected by a discomfort index, can diminish or disappear just increasing the calculation period since a larger number of occupied hours are counted, when reaching comfort is easier due to favorable outdoor conditions independently by the quality of the building.

Thus, the aforementioned long-term discomfort indices are compared in Fig. 3.4 in order to assess the influence of the expansion or compression of the calculation period.

The expansion of the calculation period, if centered on the critical part of the season, causes a significant reduction of asymmetric indices (see NaOR); on the contrary, for the symmetric indices accounting also for overcooling, it causes a significant increase of the discomfort index in the variants with the lower average indoor operative temperatures (see $POR_{Fanger,PMV}$, $POR_{Adaptive}$ and $<PPD>$); vice versa its reduction. The effect of the modification of the calculation period is more evident when discomfort values decrease. This confirms the Nicol et al.'s assertion that the expansion of the calculation period, if centered on the critical part of the season, causes a reduction of the asymmetric indices. Changing the length of the calculation period can also imply a modification of the relative evaluation between two building variants.

3.3.1.2 Displacement of the Calculation Period

The effect of the displacement of the calculation period on indices has been investigated anticipating and delaying the extremes of the reference period of 15 days. Results are reported in Fig. 3.5.

Since symmetric indices are affected also by overcooling occurrences, the anticipation of the calculation period (Test B.1) causes (i) the reduction of discomfort for those variants dominated by overheating and (ii) the increase of discomfort for those variants dominated by overcooling. The delay of the calculation period causes an increase of all discomfort performances. $POR_{Adaptive}$ is the most sensible to the displacement of the calculation period. The asymmetric index, NaOR, records a general decrease of discomfort for the anticipation of the calculation period referred to the reference period up to 55 % and a general rise to 135 % for the delay.

Finally, the anticipation of the calculation period towards spring months causes a general reduction of the discomfort due to overheating occurrences while the delay toward fall, increases the discomfort score. The variants dominated by overcooling show a rise of the predicted discomfort when the calculation period is anticipated. Shifting the period can change the relative evaluation between two variants. This is obviously relevant for an optimization process.

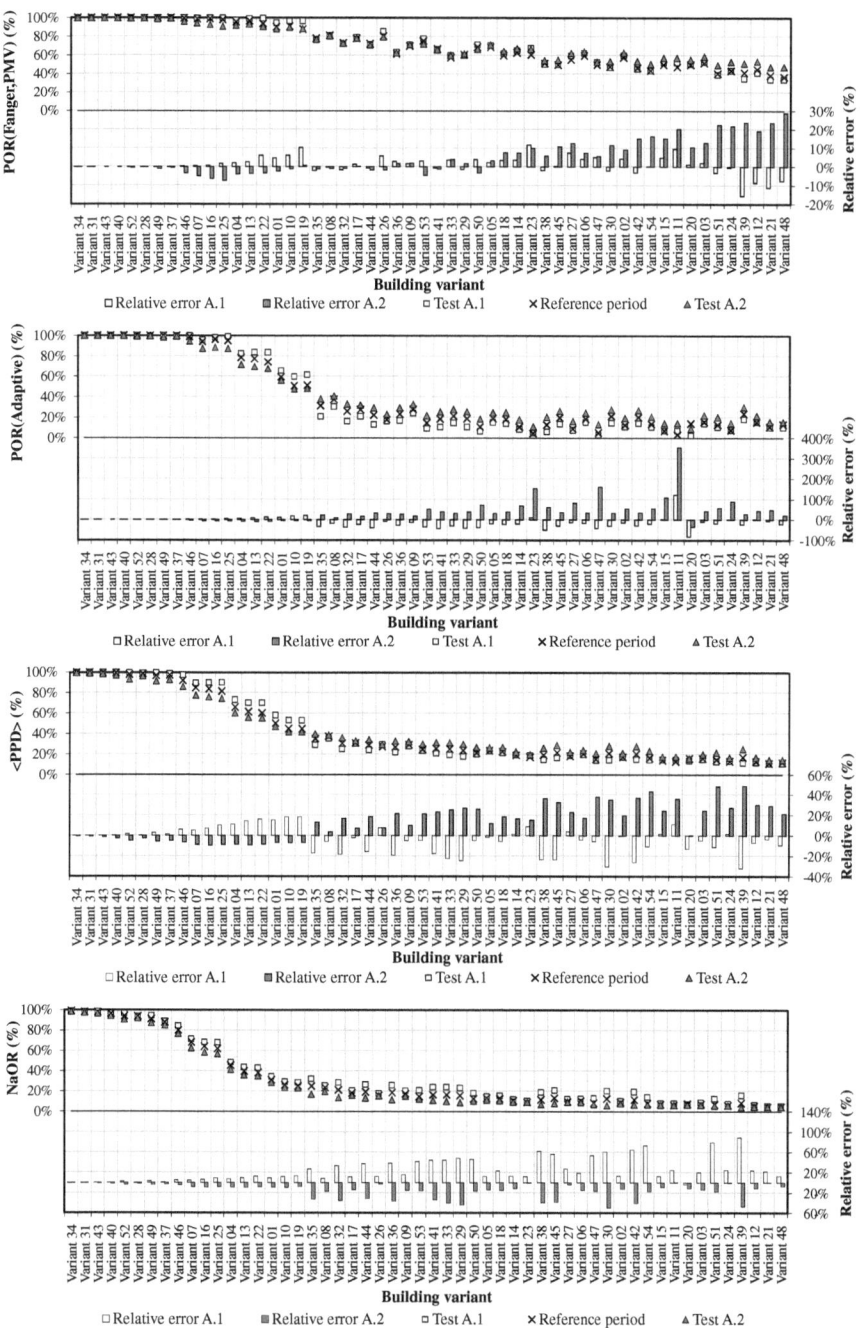

Fig. 3.4 Influence of the duration of the calculation period on long-term discomfort indices

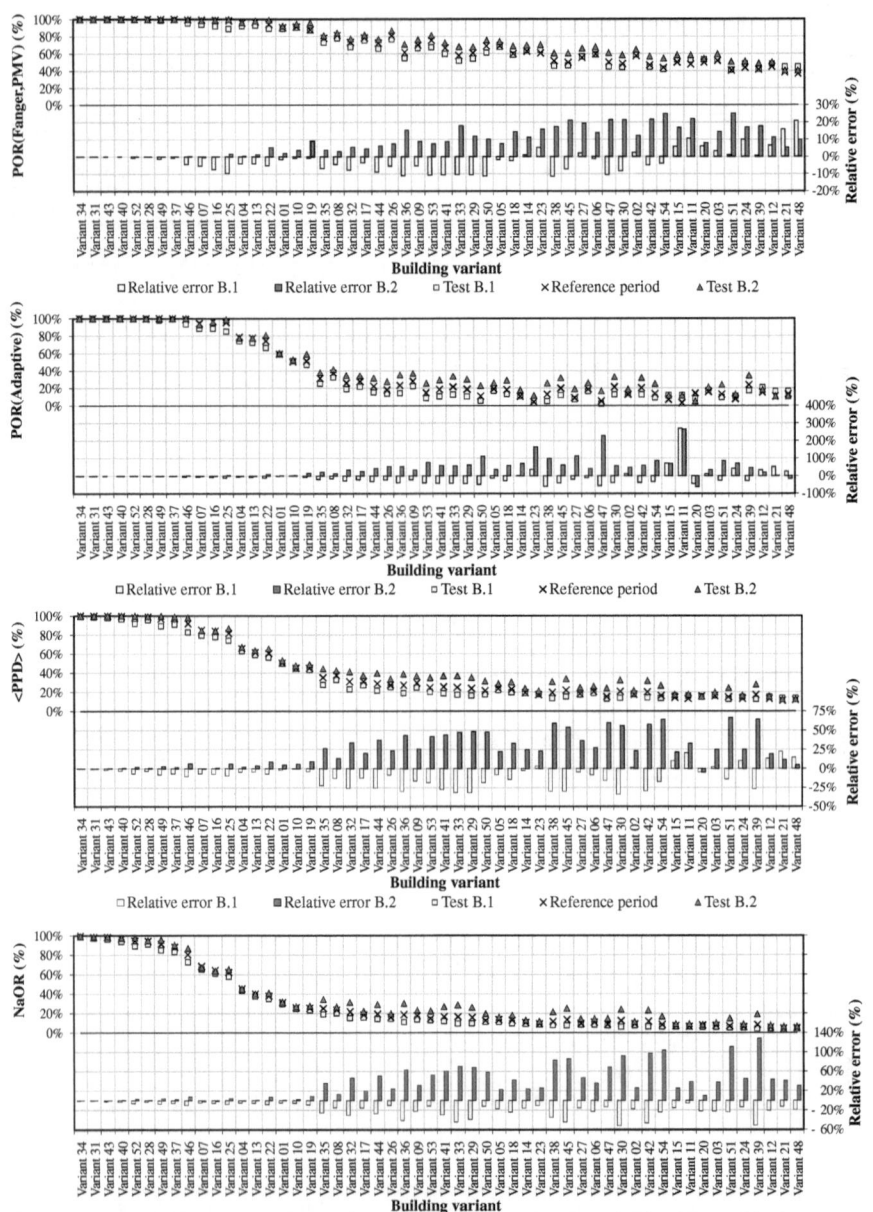

Fig. 3.5 Influence of the shift of the calculation period on a selection of the long-term discomfort indices

3.3.2 Duration of the Daily Occupation Schedule

The effect of the variation of occupancy schedule is reported in Fig. 3.6. Regarding the reference case when the building is unoccupied from 21:00 to 7:00 (9 h), two scenarios were tested:

- Scenario C.1 consists of a 2-h expansion of unoccupied time from 20:00 to 8:00 (11 h).
- Scenario C.2 consists of a 4-h expansion of unoccupied time from 19:00 to 9:00 (13 h).

The reduction of occupied time during night hours causes the increase of the asymmetric indices up to around 10 %. Thus, for these indices, an overheating problem can be solved just expanding the number of hours in the calculation schedule. For the symmetric indices, there are two possible effects: in the variants dominated by overheating occurrences, the reduction of the daily number of occupied hours cause the rising of the indices, while in the variants dominated by overcooling occurrences, discomfort decreases.

3.3.3 Weak Definitions and Simplifications in Standards

Sensitivity analyses are used to identify the dependence of the selected indices on the uncertainty of simplifications in standards: the Fanger boundary temperatures proposed by EN 15251 and the meteorological input for calculating the adaptive comfort temperature in ASHRAE 55.

3.3.4 Fanger Boundary Temperatures of Comfort Categories in EN 15251

According to ISO 7730 and EN 15251, the boundaries of comfort categories based on Fanger model are defined in terms of PMV, but they can also be expressed in terms of operative temperature. This means that boundary values of comfort categories defined in terms of PMV values can be translated into operative temperatures, by making certain assumptions about the values of external work rate, metabolic rate, clothing resistance, air velocity, relative humidity and the relationship between dry-bulb air temperature, mean radiant temperature and operative temperature.

Although EN 15251 refers to ISO 7730 for the Fanger model [17] and the PMV limits of the categories are the same (± 0.2 for Category I or A, ± 0.5 for Category II or B and ± 0.7 for Category III or C) (see Sect. 1.2), the boundary operative temperatures proposed in Table A.3 of EN 15251 are different from those reported

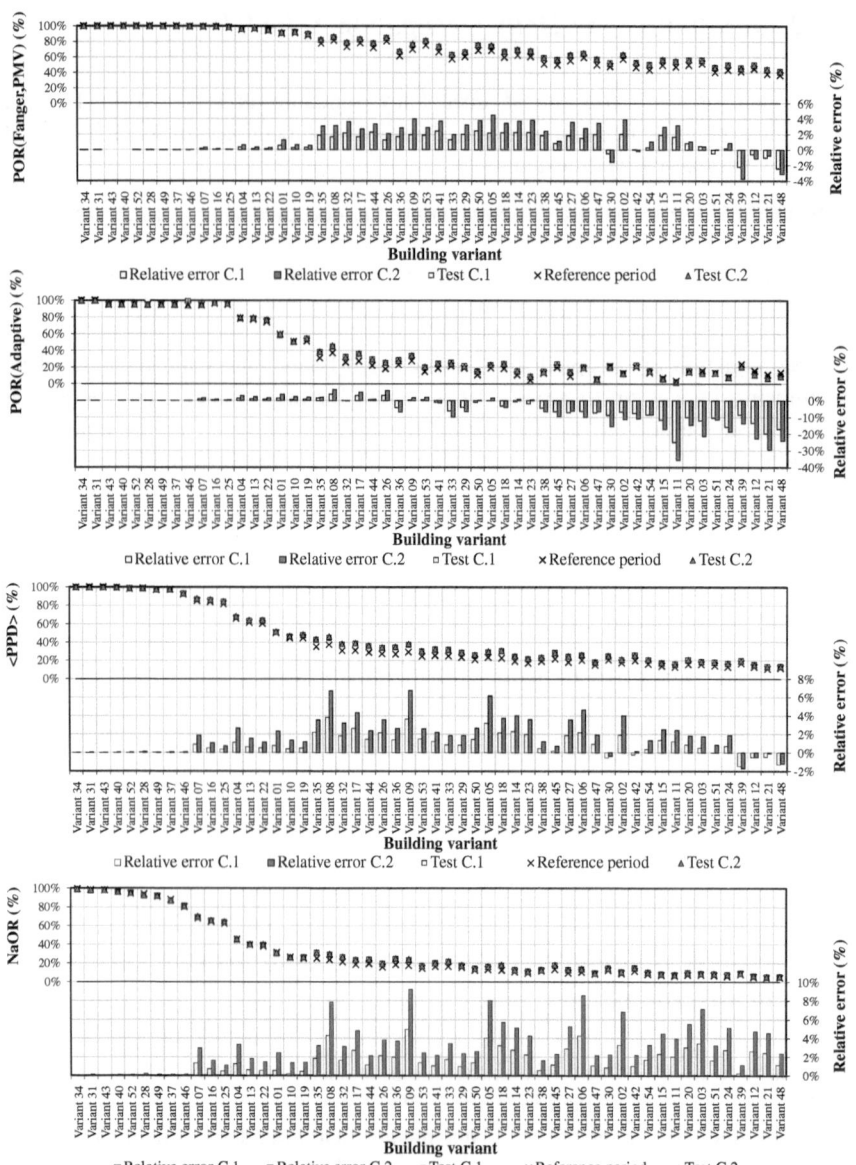

Fig. 3.6 Influence of the occupation schedule on a selection of long-term discomfort indices

in Table E.3 of ISO 7730, even if they are subject to similar assumptions: metabolic activity of 1.2 met, external work of zero met and relative humidity of 50 %.

Regarding the value of air velocity, ISO 7730 clearly indicates its value in the table while EN 15251 does not report it explicitly, but just states "low air velocity" [1].

The ISO 7730 boundary temperatures are the same of those calculable directly applying the Fanger comfort model (according to the specified assumptions) while the values proposed by EN 15251 are rounded not necessary to the nearest unit (winter conditions) or half unit (summer conditions). Also, the theoretical comfort temperatures according to EN 15251 is higher of 0.5° C in winter and lower of 0.2 °C in summer, than the ISO 7730 ones (Fig. 3.7).

The reason for this choice in EN 15251 might have been a search for simplification, i.e., to round values to tenths with accuracy of half a degree; however such simplification causes an increase of the required energy because the optimal comfort temperature and the boundary operative temperatures of every category are higher in winter and are (almost always) lower in summer.

The difference of the translation of category boundaries from PMV to operative temperature affects $POR_{Fanger,\theta op}$ and DhC (Fig. 3.8). The latter is formulated differently in the two standards; hence, we compare the two formulations calculated with the correspondent boundary temperatures.

The use of the boundary values expressed in temperature proposed by EN 15251 does not change the trends with respect to ISO 7730, but causes the increase of the value of the discomfort index up to 35 % for $POR_{Fanger,\theta op}$ and up to 60 % for DhC.

For $POR_{Fanger,op}$ the relative error increases for those building variants characterized by a lower average indoor operative temperature; on the contrary, for the DhC, the offset of the EN 15251 case from the ISO case aim at decreasing for such

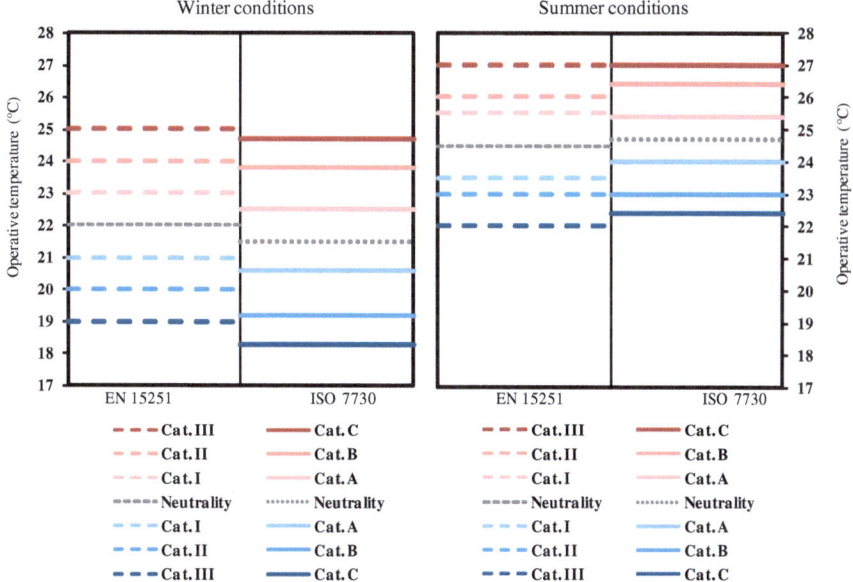

Fig. 3.7 Comparison between the Fanger boundary operative temperatures for every category proposed in ISO 7730 and EN 15251

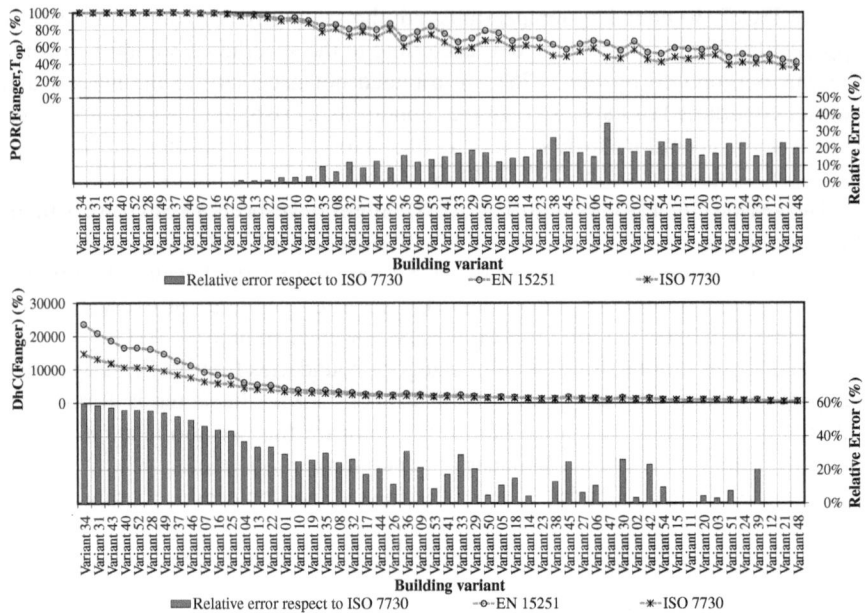

Fig. 3.8 Comparison between the indices for long-term evaluation calculated according to ISO 7730 and EN 15251 boundary values (expressed in temperature) for the comfort Category II and relative errors referred to ISO 7730

building variants. This behavior is because the indoor operative temperature exceeds more frequently the EN boundary temperatures than the ISO ones, but for very few degrees (DhC is close to zero for the coolest building variants).

3.3.4.1 Meteorological Input for the ASHRAE Adaptive Comfort Temperature

ASHRAE optimal comfort temperature is a function of the mean monthly outdoor dry-bulb air temperature. As mentioned in the literature review section, two possible interpretations about the averaging period are present in the scientific literature: the previous 30 days, and the calendar months. These two approaches cause two different curves for the ASHRAE adaptive optimal comfort temperature: a smooth function if the 30-day running average is used, on the contrary, a step function if the average over the calendar month is used (Fig. 3.9).

Exceedance$_{Adaptive}$ is the sole index that is explicitly proposed for the ASHRAE adaptive comfort model and the result of the sensitively analysis is reported in Fig. 3.10.

The impact on Exceedance$_{Adaptive}$ of the two prevailing interpretations is significant for buildings, which ensure high thermal comfort level since the magnitude of the difference due to the calculation method of the meteorological

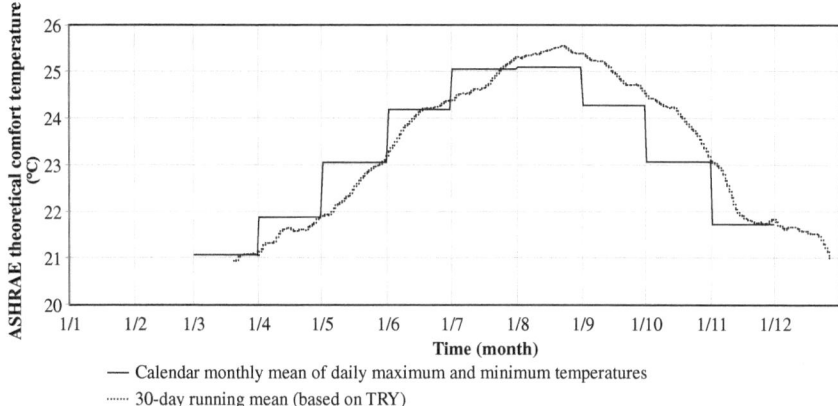

Fig. 3.9 Effect of two different definitions of monthly mean outdoor air temperature on the adaptive optimal comfort temperature according to ASHRAE 55 for the climate of Rome (IT)

input is comparable to the value of the same long-term discomfort index ($E_{Rel,-max} = 51.4\ \%$). Therefore, the gap related to the calculation method of monthly mean outdoor air temperature needs to be fixed since in high-performance buildings it could compromise the whole thermal comfort assessment of the building.

Moreover, it cannot be excluded that, if the ASHRAE adaptive comfort model is extended to calculate other indices for long-term evaluation, the effect of the changing of the calculation method of the mean monthly outdoor air temperature might be stronger because of calendar monthly averaged outdoor temperature responds rather slowly to changes in the weather and cause a step-discontinuity passing from a month to the following; on the contrary, a 30-day running mean

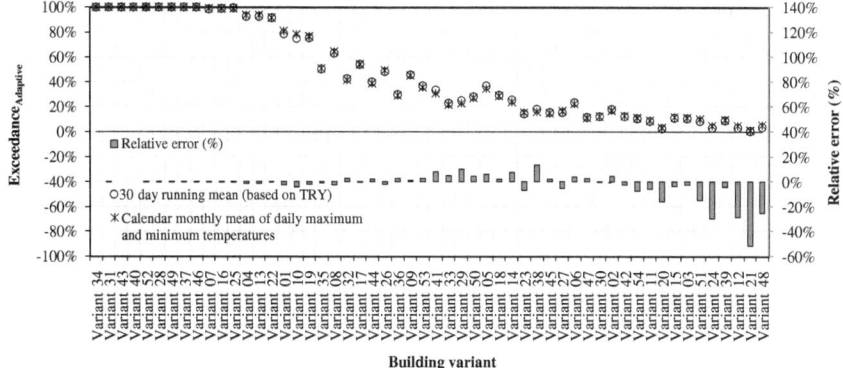

Fig. 3.10 Effect of calculation methods of mean monthly outdoor air temperature on Exceedance$_{Adaptive}$

would cause a more regular change of mean monthly outdoor air temperature and, in case of simulations, it could be simply derived from the hourly data included in the weather file, as already done by EnergyPlus.

3.4 Proposal for a Calculation Framework

In order to make the long-term discomfort indices reliable tools (i) for establishing requirements about the indoor thermal environment of buildings, (ii) for driving a comfort-optimization of buildings and (iii) for assessing their thermal comfort performance during the operational phase, the calculation of long-term discomfort indices needs to be standardized. To this end, a framework for their calculation is proposed in Table 3.2. It is an adapted version of the Table 3 of the European standard EN 15603 [18] that is instead addressed to energy calculations.

A parallel with the energy rating has been maintained since thermal comfort and energy used for the environmental control are strictly related and similar rating schemes could be simply integrated in the regulatory framework and enforce the penetration of these rating procedures in the professional field.

The thermal comfort assessment of buildings may be executed through two options, *calculated* or *measured* (the later only for existing buildings obviously). The *calculated thermal comfort rating* can be addressed to a *design* purpose, a *standardization* purpose or a *tailored* purpose:

- The *design rating* implies to refer to conventional climate, use of the building, and time schedules possibly defined at national level and it is compatible with a simplified geometrical description of the thermal zones of a building. This rating option is direct to predict the performance of a planned building in a future thermal comfort certification scheme and to support a thermal comfort optimization of a planned building.
- The *standard rating* is similar to the design rating, but it has to be applied to an existing building, and hence, it requires a description of the technical features of the building envelope and thermal plant systems as they are in the actual status of the building. This rating option is addressed for calculating the performance of an existing building in a future thermal comfort certification scheme.
- The *tailored rating*, instead, implies that climate, schedules and building data are inferred from the actual building and adapted to the purpose of the rating.
- The *measured rating* is calculated from the data metered in the actual building during its normal operation. In order to reliably assess indoor environment of a building, the measurements have to be executed in compliance with the standards in force and literature best practice about instrumentation and procedures.

Table 3.2 Proposal for a framework for the calculation of the indices for long-term evaluation of general thermal comfort conditions in a building

Rating option	Types of assessment	Thermal zoning	Input data					Utility of purpose
			Activity level	Clothing level	Occupancy and internal load	Climatic data	Building features	
Design	Calculated	Simplified	Standard	Standard	Standard	Reference year	Design	Comfort optimization, comfort certificate
Standard		Simplified	Standard	Standard	Standard	Reference year	Actual	Comfort certificate
Tailored		Depending on the purpose					Actual	Comfort optimization, validation, retrofit planning
Operational	Measured	Detailed	Actual	Actual	Actual	Actual	Actual	Comfort certificate, validation

3.4.1 Thermal Zoning

Thermal zoning can be simplified in the thermal-energy analysis of a planned building or during a *standard rating*, by aggregating those rooms which are homogeneous for exposition, type of use, time schedules, internal loads, energy emission systems and set-point temperatures [19].

3.4.2 Standard Input Parameters

Since the aim of *standard* and *design ratings* is to ensure a transversal comparison of thermal ratings of several buildings, metabolic activity, clothing resistance occupancy profiles, and lighting and plug load density and schedules should be related to the use of the building, following the suggestion of EN 15603 [18] and the example of ASHRAE 90.1 [20]. For custom purpose or specific analysis, it is possible to recur to the *tailored rating*.

3.4.3 Climatic Data

Regarding the climate, Crawley [21] suggests using the *Typical Meteorological Year*—TMY2 version based on the period of record 1961–1990 [22]—or the *Weather Year for Energy Calculations*—WYEC2 updated 2nd version [23]—for energy simulation of buildings. However, WYEC2 is available only for US locations. In 2001, the *International Weather for Energy Calculation* was released by ASHRAE [24]. The IWECs are the result of ASHRAE Research Project 1015 by *Numerical Logics and Bodycote Materials Testing Canada* for ASHRAE Technical Committee 4.2 Weather Information. The IWEC data files are *typical* weather files suitable for use with building energy simulation programs for 227 locations outside the USA and Canada. The files are derived from up to 18 years of DATSAV3 hourly weather data originally archived at the U.S. National Climatic Data Center. The weather data is supplemented by solar radiation estimated on an hourly basis from earth–sun geometry and hourly weather elements, particularly information about cloud coverage. Thus, in case of design rating, one can use IWEC or other similar sources (e.g., Meteonorm) for energy simulation of buildings. In case of tailored and measured ratings, the actual weather parameters of the period under analysis have to be collected and used.

3.4.4 Monitoring Campaign

The calculation of the thermal comfort rating shall be made for each room metered, documenting the instruments and the operational conditions of the metering. The overall rating shall be weighted for the number of people staying in each zone of the building for at least one hour in order to avoid that their prior exposure to different environmental conditions may affect their comfort perception [2]. Therefore, visitors that remain for less than one hour should not be taken into account when calculating the overall index.

3.5 Proposal for a Method for Identifying the Calculation Period

Often the terms *winter* and *summer, warm* and *cold periods, heating* and *cooling seasons* are used in the literature and the standards; but those terms are not further defined, so the analyst (being a researcher, designer or assessment officer) using the standard is left without guidance on how to choose precisely those intervals. Because of the dependence of long-term indices from duration and beginning of the calculation period, a precise definition is, in fact, necessary in order to obtain comparable results across a number of analysts. The objective is to identify an operative definition of the calculation period that is based on the relationship between outdoor conditions and the indoor thermal comfort target. A simple methodology to precisely identify the warm and cold periods is thus proposed. The main idea is to compare outdoor climatic conditions characterizing a site and the trend of the comfort temperature of a specified comfort model. For the Northern Hemisphere, the *cold period* (or *winter*) might be defined as the period when a metric based on the outdoor temperature starts to be lower than the comfort target during the second semester and ends when it becomes higher than the comfort target in the first semester. The *warm period* (or *summer*) might be defined as the period when a metric based on the outdoor temperature starts to be higher than the comfort target during the first semester, and ends when it becomes lower than the comfort target in the second semester.

The thermal effect of outdoor climate on a building is mainly due to (i) heat exchanged via natural convection by the building with external air (related to outdoor dry-bulb temperature), (ii) solar radiation incident on building envelope and (iii) heat exchanged via long-wave radiation with the surroundings and the sky. Assuming to neglect (i) the latent contribution of air humidity, (ii) the increased convective heat exchange due to wind speed, (iv) the long-wave radiation exchanged with the surroundings, the overall driving force of outdoor climate on a building can by summarized by the *Sol–air temperature,* θ_{os}, that is defined as [25]

$$\theta_{os} = \theta_{db,out} + \frac{\alpha I}{h_{c,r}} - \frac{\Delta q_i}{h_{c,r}} \tag{3.1}$$

where, $\theta_{db,out}$ is dry-bulb air temperature (°C); I is global solar irradiance on a horizontal surface (W m^{-2}), α is the solar absorbance of the surface, $h_{c,r}$ is the heat

Fig. 3.11 Comparison between 15-day mean comfort temperatures and 15-day mean sol–air temperature and identification of the warm and cold period for Palermo (IT)

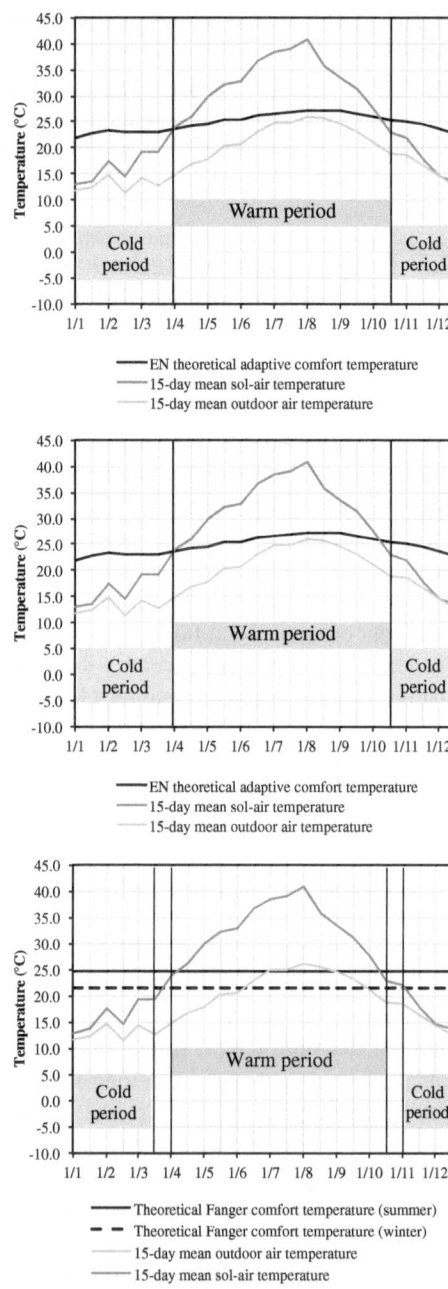

transfer coefficient for radiation and convection (W m^{-2}K^{-1}), Δq_i is a correction due to long-wave infrared radiation transferred between building surfaces and the sky. The term $\Delta q_i/h_{c,r}$ varies from close zero for vertical surfaces, to 3.9 °C for

Table 3.3 Comparison among the calculation period resulting from the three comfort models for Palermo (IT)

Comfort model	Jan		Feb		Mar		Apr		May		Jun		Jul		Aug		Sep		Oct		Nov		Dec	
	1-14	15-31	1-14	15-29	1-14	15-31	1-14	15-30	1-14	15-31	1-14	15-30	1-14	15-31	1-14	15-31	1-14	15-30	1-14	15-31	1-14	15-30	1-14	15-31
EN adaptive																								
ASHRAE adaptive																								
Fanger																								

Cold period (or Winter) Warm period (or Summer) Intermediate season

upward facing surfaces, since the sky overhead is colder than the rest of the environment. In design conditions, $\alpha/h_{c,r} = 0.052$ m^{2}K W^{-1} can be used for dark surfaces [25].

The hourly meteorological data for calculating the sol–air temperature can be derived from the weather file used in energy simulation or measured though a weather station, depending on the objective of the analysis. We suppose that 15 days is a suitable timescale resolution in order to average and compare both sol–air and the comfort temperatures. As an example, the proposed methodology has been applied to the climate of Palermo (Southern Italy) using the weather data from the IWEC file of Palermo Punta Raisi (Fig. 3.11).

The comparison between optimal comfort temperature and sol–air temperature shall be done for each 15-day bin. Resulting calculation periods are slightly different using the three comfort models (Table 3.3).

The Fanger model could admit two periods when the 15-day mean sol–air temperature is comprised between the two comfort thresholds calculated assuming a clothing resistance of 0.5 clo and 1.0 clo respectively. These two periods may be interpreted as the *intermediate seasons*.

3.6 Conclusions

The long-term discomfort indices are mathematical formulations that integrate the exceedance from an assumed thermal comfort condition over a given period. Standards and guidelines (quite) completely define the procedures for calculating such comfort conditions, commonly expressed as a range around a comfort temperature or neutrality. However, some aspects of the procedures for calculating the indices are not fully stated. In particular, three main gaps have been identified through a literature review: (i) the definition of the calculation period, (ii) the Fanger boundary temperatures as proposed in EN 15251 and (iii) the input meteorological variable for calculating the adaptive comfort temperature according to ASHRAE 55.

The effects of these gaps have been evaluated via a sensitivity analysis on the building numerical model described in Sect. 2.2.1. The main results of the sensitively analysis are:

- The precise definition of the calculation period is necessary for the long-term assessment of buildings. Such indices are sensible to the duration and to the shift of the calculation period. Its modification is more evident with the decrease of discomfort values. Changing the length of, or shifting the calculation period also implies a modification of the ranking of variants.
- The detailed reporting of the occupied schedule is crucial during operational assessment of thermal comfort condition in real buildings since increasing the daily hours of occupation can reduce an overheating problem evaluated with an asymmetric index. Standardized occupation schedules could be useful for assessing groups of buildings and necessary toward a harmonized thermal comfort certification of buildings.
- The use of the boundaries values proposed by EN 15251 does not change the trends of the indices with respect to ISO 7730, but causes the increase of the value of the discomfort index up to 35 % for $POR_{Fanger,\theta op}$ and to 60 % for DhC.
- The calculation method of monthly mean outdoor air temperature needs to be made more explicit in a further version of ASHRAE 55 since the error caused by different calculation methods of the meteorological input is comparable to the value of $Exceedance_{Adaptive}$ for those variants characterized by very low values of the index.

Then, a proposal for a harmonized calculation framework for the long-term discomfort indices is proposed. It is an adapted version of what proposed by the standard EN 15603 regarding the energy rating of buildings. The proposed scheme identifies four different typologies of thermal comfort rankings that respond to different purposes. For each ranking typology, different types of data are required, or approaches are suggested in order to limit the arbitrary analyst's decisions during the calculation phase in order to make standard and design rankings comparable among different buildings. This calculation framework could be considered the first attempt to set a harmonized and homogeneous thermal comfort certification scheme that could complete the current energy certification schemes, e.g., derived from EN 15603.

Finally, a method for defining the duration and the beginning of the seasons with respect to the comfort targets set in a specified building is presented in order to allow comparable rankings of buildings sited in different climatic conditions and to limit arbitrary analyst's choices that influence significantly the results.

References

1. CEN: EN 15251, *Indoor Environmental Input Parameters for Design and Assessment of Energy Performance of Buildings Addressing Indoor Air Quality, Thermal Environment, Lighting and Acoustics* (European Committee for Standardization, Brussels, 2007)
2. ASHRAE: ANSI/ASHRAE 55, *Thermal Environmental Conditions for Human Occupancy* (American Society of Heating, Refrigerating and Air-Conditioning Engineers, Atlanta, 2004)

3. ISO: ISO 7730, *Ergonomics of the Thermal Environment—Analytical Determination and Interpretation of Thermal Comfort Using Calculation of the PMV and PPD Indices and Local Thermal Comfort Criteria*, 3rd version. (International Standard Organization, Geneva, 2005)

4. CIBSE: Guide J, *Weather, Solar and Illuminance Data* (Chartered Institution of Building Services Engineers, London, 2002)

5. CIBSE: Guide A, *Environmental Design* (Chartered Institution of Building Services Engineers, London, 2006)

6. J.F. Nicol, J. Hacker, B. Spires, H. Davies, Suggestion for new approach to overheating diagnostic. in *Proceedings of Conference: Air Conditioning and the Low Carbon Cooling Challenge,* Cumberland Lodge, Windsor (2008)

7. AMS, Statement on seasonal to interannual climate prediction (Adopted by AMS Council 14 January 2001). Bull. Am. Meteorol. Soc. **82**, 701 (2001)

8. P. Alpert, I. Osetinsky, B. Ziv, H. Shafir, A new seasons definition based on classified daily synoptic systems: an example for the Eastern Mediterranean. Int. J. Climatol. **24**, 1013–1021 (2004)

9. K.E. Trenberth, What are the seasons? Bull. Am. Meteorol. Soc. **64**, 1276–1282 (1983)

10. R.J. de Dear, G.S. Brager, Towards an adaptive model of thermal comfort and preference. ASHRAE Trans. **104**(1), 145–167 (1998)

11. B.W. Olesen, O. Seppanen, A. Boerstra, Criteria for the indoor environment for energy performance of buildings—a new European standard. Facilities **24**, 445–457 (2006)

12. F. Haldi, D. Robinson, On the behaviour and adaptation of office occupants. Build. Environ. **43**, 2163–2177 (2008)

13. D.E. Kalz, J. Pfafferott, Comparative Evaluation of Natural Ventilated and Mechanical Cooled Non-Residential Buildings in Germany: Thermal Comfort in Summer. in *Proceedings of Conference: Adapting to Change: New Thinking on Comfort*, Cumberland Lodge, Windsor (2010)

14. S. Borgeson, G.S. Brager, Comfort Exceedance Metrics in Mixed-Mode buildings. in *Proceedings of Conference: Adapting to Change: New Thinking on Comfort* (Cumberland Lodge, Windsor, 2010)

15. C. Huizenga, *ASHRAE 1332-RP Revisions to the ASHRAE thermal comfort tool to maintain consistency with Standard 55:2004* (American Society of Heating, Refrigerating and Air-Conditioning Engineers, Atlanta, 2004)

16. US DOE: EnergyPlus Engineering Reference, *The Reference to EnergyPlus Calculations* (version 7), U.S. Department of Energy (2012)

17. P.O. Fanger, *Thermal Comfort* (Danish Technical Press, Copenhagen, 1970)

18. CEN: EN 15603, *Energy Performance of Buildings—Overall Energy Use and Definition of Energy Ratings* (European Committee for Standardization, Brussels, 2008)

19. CEN: EN 13790, *Energy Performance of Buildings—Calculation of Energy Use for Space Heating and Cooling* (European Committee for Standardization, Brussels, 2008)

20. ASHRAE: ANSI/ASHRAE/IES Standard 90.1, *Applicability to Datacom* (American Society of Heating, Refrigerating and Air-Conditioning Engineers, Atlanta, 2010)

21. D.B. Crawley, Which weather data should you use for energy simulations of commercial buildings? ASHRAE Trans. **104**(2), 1–18 (1998)

22. NREL: User's Manual for TMY2s (Typical Meteorological Years), NREL/SP-463-7668, and TMY2s, Typical Meteorological Years Derived from the 1961–1990 National Solar Radiation Data Base, National Renewable Energy Laboratory (NREL), CDROM. Golden (1995)

23. ASHRAE: WYEC2 Weather Year for Energy Calculations 2, *Toolkit and Data* (American Society of Heating, Refrigerating and Air-Conditioning Engineers, Atlanta, 1997)

24. ASHRAE: International Weather for Energy Calculations (IWEC Weather Files), *Users Manual and CD-ROM* (American Society of Heating, Refrigerating and Air-Conditioning Engineers, Atlanta, 2001)

25. J.F. Kreider, P.S. Curtiss, A. Rabl, *Heating and Cooling of Buildings: Design for Efficiency*, Revised 2nd edn. (Taylor & Francis, Boca Raton, 2010)

Chapter 4
The Long-Term Percentage of Dissatisfied

Abstract Following the analysis about the indices for the long-term comfort evaluation, carried out in the previous two chapters, in this chapter a new long-term discomfort index, called *Long-term Percentage of Dissatisfied*, is proposed, and its features are discussed. It is a symmetric and comfort-model based index, not depending on comfort categories, and it is applicable to both summer and winter assessments. This index is proposed in three versions, which differ in the way the hourly likelihood of thermal discomfort is evaluated, depending on the reference comfort model used for the assessment. The three versions weigh the zone hourly likelihood of thermal discomfort by the hourly occupancy rate in each thermal zone of the building. The first version of the new index is based on the Fanger comfort model, and its likelihood of thermal discomfort coincides with the Predicted percentage of dissatisfied. The second version is based on the EN adaptive comfort model, and its likelihood of thermal discomfort is calculated through the Nicol et al.'s overheating risk. The third is based on the ASHRAE adaptive comfort model, and, since a reliable likelihood of thermal discomfort was not identified in literature for this comfort model, it has been developed via a logistic regression analysis of the data collected in the ASHRAE RP-884 database. Therefore, the methodology adopted for deriving such new likelihood of thermal discomfort from the ASHRAE RP-884 data, and the result of the performed logistic regression analysis are presented. Then, the three versions of the index are calculated for the 54 building variants described in Chap. 1 and their results are compared. Finally, since the building performance simulation software used in this work, EnergyPlus, does not directly calculate the proposed new index, three computer codes have been written in the programming language EnergyPlus Runtime Language, and then integrated in EnergyPlus through the Energy Management System module.

S. Carlucci, *Thermal Comfort Assessment of Buildings*,
PoliMI SpringerBriefs, DOI: 10.1007/978-88-470-5238-3_4,
© The Author(s) 2013

4.1 Introduction

In Chaps. 1 and 2, a review and a comparison of a number of long-term discomfort indices are proposed, and strengths and weaknesses of such indices have been identified. The overall result is that none of them seems to be completely reliable for the long-term evaluation of the general thermal comfort conditions in buildings, but most of them include appealing features that can serve as inspiration in developing a new index. At the same time, the rule proposed in the standards ISO 7730 [1] and EN 15251 [2] for averaging the indices over the net floor area of thermal zones in multi-zone buildings shows some weaknesses since it does not account for the amount of people inside every zone, but dilutes the indices over the extension of the zones. Moreover, the binary code commonly used for evaluating occupancy schedule can cause an unsuitable hard cutoff in those zones where occupancy gradually changes.

In this chapter, a new index for the long-term evaluation of the general comfort conditions in buildings is proposed in order to satisfy all desired features identified in the previous chapters.

4.2 A New Long-Term Discomfort Index

Taking inspiration form the conclusions of Chaps. 1 and 2, a reliable new long-term discomfort index should:

- Be applicable for both free-running and mechanically cooled buildings, and should be suitable for being used with both the adaptive and Fanger comfort models (e.g., similarly to Percentage outside range indices).
- Reflect the nonlinear relationship between thermal discomfort and the exceedance from the theoretical optimal comfort conditions (e.g. similarly to Nicol et al.'s overheating risk and Average PPD).
- Evaluate thermal discomfort by putting more focus on the occupants, rather than on building geometry (e.g., similarly to Exceedance indices).
- Be applicable to the evaluation of uncomfortable conditions during both summer and winter.
- Allow estimating separately possible uncomfortable conditions due to upper and lower exceedance from theoretical comfort temperature.
- Be independent of comfort categories.

Moreover, in order to guarantee a reliable thermal comfort assessment, a number of boundary conditions must be clearly specified:

- The standard used for the derivation of the operative temperature boundary from PMV limits (ISO 7730 or EN 15251), for those indices that use operative temperatures derived from the Fanger model.

- The method used to calculate the meteorological input data for the ASHRAE optimal comfort temperature in naturally ventilated buildings (monthly or 30-day running mean outdoor air temperature).
- The starting and ending days of the calculation period.

In order to satisfy all aforementioned features, a weighted percentage index is proposed. The basic idea is to build an index that summarizes the likelihood of thermal discomfort globally perceived by all occupants (real or hypothetical) of all zones in a building during a certain season, in a single percentage value. Hence, the problem is split into three sub-problems: (i) the estimation of the likelihood of thermal discomfort perceived by one occupant, (ii) the averaging method of every zone indices in a multi-zone building, and (iii) the integrations over occupation and calculation periods.

4.2.1 The Likelihood of Thermal Discomfort at a Specified Time and Space

An index for the long-term evaluation should estimate the likelihood of thermal discomfort perceived by a typical occupant. Thus, reference is made to the thermal comfort models, as well as they have been defined in the standards (Fanger in ISO 7730, the EU adaptive in EN 15251 and American adaptive in ASHRAE 55 [3]) and for each of them, a mathematical relationship has been identified in order to estimate the severity of the deviations from a theoretical thermal comfort objective, given certain outdoor and indoor conditions at specified time and space location. The hourly timescale resolution is considered more appropriate for its assessment and such mathematical relationships are denoted in this model with the term Likelihood of Dissatisfied (LD).

One aspect not yet fully covered in the scientific literature is the phenomenon of overcooling in summer [4]. It indicates a condition when the indoor operative temperature is lower than the theoretical comfort value according to the chosen comfort model.

The assumption that the distribution of thermal sensation votes is symmetric with respect to the neutral term derives from the PMV/PPD relation proposed by Fanger [5]. de Dear et al. [6] also used such relationship; furthermore, they assumed linearity between the thermal sensation votes and indoor operative temperatures. Thus, combining the two assumptions, results that also the temperature distribution is symmetric with respect to the neutral temperature. This assumption is also used implicitly in the definition of thermal comfort categories that are equally spaced with respect to the theoretical comfort temperature. The same assumption is adopted in this research work.

The categories are defined in the standards as discrete intervals of a chosen variable (PMV or operative temperature), and the choice of these intervals is, to some degree, a convention and is the subject of a debate. Also, since long-term discomfort indices could be used to drive a comfort-optimization process, the use of comfort categories within the definition of the objective function (e.g., as it happens in the framework of EN 15251) would introduce arbitrary step changes and would also disturb the optimization algorithms; for these reasons, the aim is to construct an analytically *smooth* function.

For the Fanger comfort model, the index that estimates the likelihood of thermal discomfort perceived by a typical occupant is the Predicted Percentage of Dissatisfied (PPD) which is directly computable from the Predicted Mean Vote (PMV), using the equation

$$PPD^{ISO}_{Fanger}(PMV) = 100 - 95e^{-0.03353 \cdot PMV^4 - 0,2179 \cdot PMV^2} \in [5, 100]. \qquad (4.1)$$

Regarding the EN Adaptive model, the Nicol et al.'s overheating risk (NaOR) [7] was considered adequate to estimate the likelihood of thermal discomfort. It was derived through a logistic regression from the thermal sensation votes collected in real office buildings in free-floating mode

$$P^{EN}_{Adaptive}(\Delta\theta_{op}) = NaOR \equiv \frac{e^{0.4734 \cdot \Delta\theta_{op} - 2.607}}{1 + e^{0.4734 \cdot \Delta\theta_{op} - 2.607}} \in [0.07, 1.00]. \qquad (4.2)$$

where $\Delta\theta_{op} = |\theta_{op,in} - \theta_{comf,EN}|$. The NaOR was proposed to assess exclusively summer overheating in free-floating buildings; in order to construct a complete function to represent the likelihood of thermal discomfort, we use it to assess symmetrically not only overheating, but also overcooling.

Regarding the ASHRAE adaptive model, an index for estimating the likelihood of thermal discomfort is missing in literature; thus it was necessary to build a new analytical function, called *ASHRAE Likelihood of Dissatisfied (ALD)*, which has been determined via a logistic regression analysis performed on the data collected in the ASHRAE RP-884 database. Its derivation is described in detail in the next section. Its final expression is

$$P^{ASHRAE}_{Adaptive}(\Delta\theta_{op}) = ALD$$

$$ALD \equiv \frac{e^{0.008 \cdot \Delta\theta_{op}^2 + 0.406 \cdot \Delta\theta_{op} - 3.050}}{1 + e^{0.008 \cdot \Delta\theta_{op}^2 + 0.406 \cdot \Delta\theta_{op} - 3.050}} \in [0.05, 1.00]. \qquad (4.3)$$

where $\Delta\theta_{op} = |\theta_{op,in} - \theta_{comf,ASHRAE}|$.

In order to provide a graphical comparison among the three selected likelihood distributions, it is necessary to present them in a form where they have the same input variable, e.g., the operative temperature. Hence, for the distribution based on the Fanger model, a selection of PMV values have to be translated in operative temperatures by assuming, as done in Chap. 2, that the dry-bulb air temperature is equal to the mean radiant temperature (hence equal to the operative temperature), indoor air relative humidity of 50 %, air velocity of 0.1 m s^{-1}, metabolic activity of 1.2 met, clothing resistance of 0.5 clo in summer conditions, and an external work of zero met. According to this, the likelihood distributions derivable for the Fanger model are two: one for summer and one for winter.

Operative temperatures and PPDs are calculated every 0.2 PMV steps in the interval from −3.0 to + 3.0 (Table 4.1).

The pairs (offset from optimal comfort temperature and PPD) have been also calculated with the same procedure for the winter case, when the design value of the clothing resistance is assumed 1.0 clo.

The four likelihood-of-dissatisfied distributions are represented in Fig. 4.1.

They show quite similar trends, although there are a number of differences:

- Their scopes: the Fanger ones are addressed to mechanically cooled buildings and the adaptive ones are addressed to naturally ventilated/cooled buildings.
- The datasets from which they were derived: data recorded in thermal chamber [5], SCATs Project [8] and ASHRAE RP-884 [6].
- The methods used for deriving them: Griffiths's method [9] and binning of sample data method [6].
- The values of their theoretical comfort temperatures.

4.2.2 Averaging Over Different Zones of a Building

In the case of a multi-zone building, and assuming that the thermal conditions are homogeneous in each zone, a value of a given long-term discomfort index has to be calculated for each zone of the building. Therefore, in order to represent the performance of the whole building into a single value, an average of such zone indices has to be calculated. The EN 15251 proposes to weigh by a geometrical feature of the building (in two different sections it refers to the net floor area and to net volume). However, the indices proposed by ISO 7730 (excepted the Average PPD and the Sum of PPD) and EN 15251 are mathematical models built upon the comfort models, which measure the frequency with which certain boundary conditions are exceeded (e.g., POR), or accumulate hours when the indoor parameters are above or below specified boundary conditions (e.g., DhC and PPDwC) (see Chap. 1);

On the contrary, if a long-term discomfort index measures the likelihood of thermal discomfort perceived by occupants in a certain thermal zone (<PPD>, NaOR and ALD), it seems to be more appropriate to weigh each zone index for the number of people occupying the specified zone at a certain hour, rather than to weigh it over the zone floor area or volume.

Data about zone occupancy may be (i) deduced in existing buildings by monitoring of certain physical parameters directly related to the number of occupants of the zone (e.g., the concentration of carbon dioxide in the air), or (ii) assumed based on information about building use or inserted in the time schedule filled in a dynamic simulation. This would overcome the binary-step function that characterizes the majority of the long-term discomfort indices: the

Table 4.1 Derivation of the pairs with offset from thermal neutrality temperature and PPD (summer conditions)

PMV	θ_{op} (°C)	$\Delta\theta_{op}$(°C) Z	PPD (%)	PMV	θ_{op} (°C)	$\Delta\theta_{op}$(°C) Z	PPD (%)
-3.0	15.0	-9.8	99.1	0.0	24.7	0.0	5.0
-2.8	15.7	-9.1	97.8	+0.2	25.4	0.7	5.8
-2.6	16.3	-8.4	95.3	+0.4	26.1	1.3	8.3
-2.4	16.9	-7.8	91.1	+0.6	26.7	2.0	12.5
-2.2	17.6	-7.2	84.8	+0.8	27.4	2.7	18.4
-2.0	18.2	-6.5	76.7	+1.0	28.0	3.3	26.1
-1.8	18.8	-5.9	66.9	+1.2	28.7	4.0	35.2
-1.6	19.5	-5.3	56.2	+1.4	29.3	4.6	45.4
-1.4	20.1	-4.6	45.4	+1.6	30.0	5.3	56.2
-1.2	20.8	-4.0	35.2	+1.8	30.6	5.9	66.9
-1.0	21.4	-3.3	26.1	+2.0	31.3	6.5	76.7
-0.8	22.1	-2.7	18.4	+2.2	31.9	7.2	84.8
-0.6	22.7	-2.0	12.5	+2.4	32.5	7.8	91.1
-0.4	23.4	-1.3	8.3	+2.6	33.2	8.4	95.3
-0.2	24.1	-0.7	5.8	+2.8	33.8	9.1	97.8
0.0	24.7	0.0	5.0	+3.0	34.4	9.7	99.1

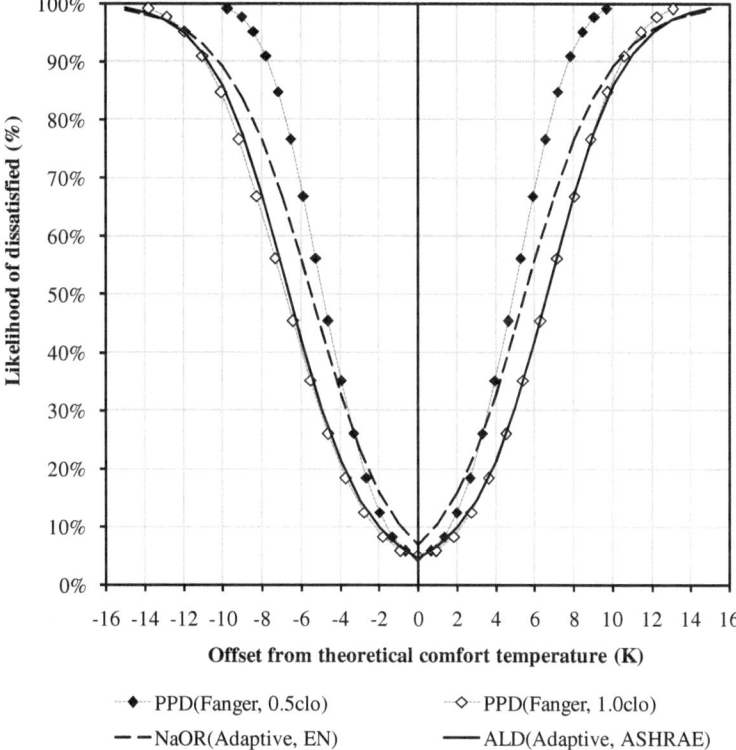

Fig. 4.1 Likelihood of Dissatisfied as a function of the offset from theoretical comfort temperature, according to the three thermal comfort models

time step weighs 1 when the zone is occupied whilst it weighs 0 when the zone is unoccupied.

The Exceedance indices [10] weigh the hours when the boundary conditions corresponding to an acceptability of 80 % are exceeded for the hourly occupation rate with respect to the total number of people inside the building, but do not specify how to average in a multi-zone building.

The proposed long-term discomfort index is weighted by the number of people occupying a given zone of the building for at least one hour. Visitors that remain for less than one hour are not taken into account when calculating the overall index.

4.2.3 Proposal for a New Long-Term Thermal Discomfort Index

The new and ameliorated index for the long-term evaluation of the general comfort conditions in a building is called *Long-term Percentage of Dissatisfied* (LPD). It is a symmetric and comfort-model based index, not depending on comfort categories, and it is applicable to both summer and winter assessments. Its analytical expression accounts for the hourly-predicted Likelihood of Dissatisfied calculated for each zone, which is weighted for the number of people inside the zone, and is normalized over the total number of the people inside the building and over all time corresponding to the seasonal calculation period

$$LPD(p, LD) \equiv \frac{\sum_{t=1}^{T} \sum_{z=1}^{Z} \left(p_{z,t} \cdot LD_{z,t} \cdot h_t \right)}{\sum_{t=1}^{T} \sum_{z=1}^{Z} \left(p_{z,t} \cdot h_t \right)} \tag{4.4}$$

where t is the counter for the time step of the calculation period, T is the last progressive time step of the calculation period, z is the counter for the zones of a building, Z is the total number of the zones, $p_{z,t}$ is the zone occupation rate at a certain time step, $LD_{z,t}$ is the Likelihood of Dissatisfied inside a certain zone at a certain time step and h_t is the duration of a calculation time step (by default one hour). This index can be used to quantify thermal discomfort during the cold period, the warm period, or the whole year, due to overheating, overcooling or both, and making reference to the Fanger model, the European adaptive model and the American adaptive model. Its formulations depend on (i) the comfort model (ii) the calculation period and (iii) the discomfort type (Tables 4.2, 4.3, 4.4).

Table 4.2 Formulations of the Long-term Percentage of Dissatisfied (Fanger)

Calculation period	Discomfort assessment	Long-term Percentage of Dissatisfied in the Fanger version, LPD^{Fanger} (%)
Warm period	Overheating	$LPD_{W,OH}^{Fanger}\big|_{PMV_{actual}>0} = \dfrac{\sum_{t=1}^{W} \sum_{z=1}^{Z} \left(p_{z,t} \cdot <PPD>_{z,t} \cdot h_t \right)}{\sum_{t=1}^{W} \sum_{z=1}^{Z} \left(p_{z,t} \cdot h_t \right)}$
	Overcooling	$LPD_{W,OC}^{Fanger}\big|_{PMV_{actual}<0} = \dfrac{\sum_{t=1}^{W} \sum_{z=1}^{Z} \left(p_{z,t} \cdot <PPD>_{z,t} \cdot h_t \right)}{\sum_{t=1}^{W} \sum_{z=1}^{Z} \left(p_{z,t} \cdot h_t \right)}$
	Overall	$LPD_{W}^{Fanger}\big|_{PMV_{actual}\neq0} = \dfrac{\sum_{t=1}^{W} \sum_{z=1}^{Z} \left(p_{z,t} \cdot <PPD>_{z,t} \cdot h_t \right)}{\sum_{t=1}^{W} \sum_{z=1}^{Z} \left(p_{z,t} \cdot h_t \right)}$
Cold period	Overheating	$LPD_{C,OH}^{Fanger}\big|_{PMV_{actual}>0} = \dfrac{\sum_{t=1}^{C} \sum_{z=1}^{Z} \left(p_{z,t} \cdot <PPD>_{z,t} \cdot h_t \right)}{\sum_{t=1}^{C} \sum_{z=1}^{Z} \left(p_{z,t} \cdot h_t \right)}$
	Overcooling	$LPD_{C,OC}^{Fanger}\big|_{PMV_{actual}<0} = \dfrac{\sum_{t=1}^{C} \sum_{z=1}^{Z} \left(p_{z,t} \cdot <PPD>_{z,t} \cdot h_t \right)}{\sum_{t=1}^{C} \sum_{z=1}^{Z} \left(p_{z,t} \cdot h_t \right)}$
	Overall	$LPD_{C}^{Fanger}\big|_{PMV_{actual}\neq0} = \dfrac{\sum_{t=1}^{C} \sum_{z=1}^{Z} \left(p_{z,t} \cdot <PPD>_{z,t} \cdot h_t \right)}{\sum_{t=1}^{C} \sum_{z=1}^{Z} \left(p_{z,t} \cdot h_t \right)}$
Whole year	Overheating	$LPD_{OH}^{Fanger}\big|_{PMV_{actual}>0} = \dfrac{\sum_{t=1}^{Y} \sum_{z=1}^{Z} \left(p_{z,t} \cdot <PPD>_{z,t} \cdot h_t \right)}{\sum_{t=1}^{Y} \sum_{z=1}^{Z} \left(p_{z,t} \cdot h_t \right)}$
	Overcooling	$LPD_{OC}^{Fanger}\big|_{PMV_{actual}<0} = \dfrac{\sum_{t=1}^{Y} \sum_{z=1}^{Z} \left(p_{z,t} \cdot <PPD>_{z,t} \cdot h_t \right)}{\sum_{t=1}^{Y} \sum_{z=1}^{Z} \left(p_{z,t} \cdot h_t \right)}$
	Overall	$LPD^{Fanger}\big|_{PMV_{actual}\neq0} = \dfrac{\sum_{t=1}^{Y} \sum_{z=1}^{Z} \left(p_{z,t} \cdot <PPD>_{z,t} \cdot h_t \right)}{\sum_{t=1}^{Y} \sum_{z=1}^{Z} \left(p_{z,t} \cdot h_t \right)}$

Table 4.3 Formulations of the Long-term Percentage of Dissatisfied (EN adaptive)

Calculation period	Discomfort assessment	Long-term Percentage of Dissatisfied in the EN adaptive version, LPD$^{EN\ Adaptive}$ (%)
Warm period	Overheating	$LPD_{W,OH}^{EN\ Adaptive}\Big|_{\theta_{op,in} > \theta_{comf}^{EN15251}} = \dfrac{\sum_{t=1}^{W}\sum_{z=1}^{Z}\left(p_{z,t}\cdot NaOR_{z,t}\cdot h_t\right)}{\sum_{t=1}^{W}\sum_{z=1}^{Z}\left(p_{z,t}\cdot h_t\right)}$
	Overcooling	$LPD_{W,OC}^{EN\ Adaptive}\Big|_{\theta_{op,in} < \theta_{comf}^{EN15251}} = \dfrac{\sum_{t=1}^{W}\sum_{z=1}^{Z}\left(p_{z,t}\cdot NaOR_{z,t}\cdot h_t\right)}{\sum_{t=1}^{W}\sum_{z=1}^{Z}\left(p_{z,t}\cdot h_t\right)}$
	Overall	$LPD_{W}^{EN\ Adaptive}\Big|_{\theta_{op,in} \neq \theta_{comf}^{EN15251}} = \dfrac{\sum_{t=1}^{W}\sum_{z=1}^{Z}\left(p_{z,t}\cdot NaOR_{z,t}\cdot h_t\right)}{\sum_{t=1}^{W}\sum_{z=1}^{Z}\left(p_{z,t}\cdot h_t\right)}$
Cold period	Overheating	$LPD_{C,OH}^{EN\ Adaptive}\Big|_{\theta_{op,in} > \theta_{comf}^{EN15251}} = \dfrac{\sum_{t=1}^{C}\sum_{z=1}^{Z}\left(p_{z,t}\cdot NaOR_{z,t}\cdot h_t\right)}{\sum_{t=1}^{C}\sum_{z=1}^{Z}\left(p_{z,t}\cdot h_t\right)}$
	Overcooling	$LPD_{C,OC}^{EN\ Adaptive}\Big|_{\theta_{op,in} < \theta_{comf}^{EN15251}} = \dfrac{\sum_{t=1}^{C}\sum_{z=1}^{Z}\left(p_{z,t}\cdot NaOR_{z,t}\cdot h_t\right)}{\sum_{t=1}^{C}\sum_{z=1}^{Z}\left(p_{z,t}\cdot h_t\right)}$
	Overall	$LPD_{C}^{EN\ Adaptive}\Big|_{\theta_{op,in} \neq \theta_{comf}^{EN15251}} = \dfrac{\sum_{t=1}^{C}\sum_{z=1}^{Z}\left(p_{z,t}\cdot NaOR_{z,t}\cdot h_t\right)}{\sum_{t=1}^{C}\sum_{z=1}^{Z}\left(p_{z,t}\cdot h_t\right)}$
Whole year	Overheating	$LPD_{OH}^{EN\ Adaptive}\Big|_{\theta_{op,in} > \theta_{comf}^{EN15251}} = \dfrac{\sum_{t=1}^{Y}\sum_{z=1}^{Z}\left(p_{z,t}\cdot NaOR_{z,t}\cdot h_t\right)}{\sum_{t=1}^{Y}\sum_{z=1}^{Z}\left(p_{z,t}\cdot h_t\right)}$
	Overcooling	$LPD_{OC}^{EN\ Adaptive}\Big|_{\theta_{op,in} < \theta_{comf}^{EN15251}} = \dfrac{\sum_{t=1}^{Y}\sum_{z=1}^{Z}\left(p_{z,t}\cdot NaOR_{z,t}\cdot h_t\right)}{\sum_{t=1}^{Y}\sum_{z=1}^{Z}\left(p_{z,t}\cdot h_t\right)}$
	Overall	$LPD^{EN\ Adaptive}\Big|_{\theta_{op,in} \neq \theta_{comf}^{EN15251}} = \dfrac{\sum_{t=1}^{Y}\sum_{z=1}^{Z}\left(p_{z,t}\cdot NaOR_{z,t}\cdot h_t\right)}{\sum_{t=1}^{Y}\sum_{z=1}^{Z}\left(p_{z,t}\cdot h_t\right)}$

Table 4.4 Formulations of the Long-term Percentage of Dissatisfied (ASHRAE adaptive)

Calculation period	Discomfort assessment	Long-term Percentage of Dissatisfied in the ASHRAE Adaptive version, LPD$^{ASHRAE\ Adaptive}$ (%)
Warm period	Overheating	$LPD_{W,OH}^{ASHRAE\ Adaptive}\Big|_{\theta_{op,in} > \theta_{comf}^{ASHRAE55}} = \dfrac{\sum_{t=1}^{W}\sum_{z=1}^{Z}\left(p_{z,t}\cdot ALD_{z,t}\cdot h_t\right)}{\sum_{t=1}^{W}\sum_{z=1}^{Z}\left(p_{z,t}\cdot h_t\right)}$
	Overcooling	$LPD_{W,OC}^{ASHRAE\ Adaptive}\Big|_{\theta_{op,in} < \theta_{comf}^{ASHRAE55}} = \dfrac{\sum_{t=1}^{W}\sum_{z=1}^{Z}\left(p_{z,t}\cdot ALD_{z,t}\cdot h_t\right)}{\sum_{t=1}^{W}\sum_{z=1}^{Z}\left(p_{z,t}\cdot h_t\right)}$
	Overall	$LPD_{W}^{ASHRAE\ Adaptive}\Big|_{\theta_{op,in} \neq \theta_{comf}^{ASHRAE55}} = \dfrac{\sum_{t=1}^{W}\sum_{z=1}^{Z}\left(p_{z,t}\cdot ALD_{z,t}\cdot h_t\right)}{\sum_{t=1}^{W}\sum_{z=1}^{Z}\left(p_{z,t}\cdot h_t\right)}$
Cold period	Overheating	$LPD_{C,OH}^{ASHRAE\ Adaptive}\Big|_{\theta_{op,in} > \theta_{comf}^{ASHRAE55}} = \dfrac{\sum_{t=1}^{C}\sum_{z=1}^{Z}\left(p_{z,t}\cdot ALD_{z,t}\cdot h_t\right)}{\sum_{t=1}^{C}\sum_{z=1}^{Z}\left(p_{z,t}\cdot h_t\right)}$
	Overcooling	$LPD_{C,OC}^{ASHRAE\ Adaptive}\Big|_{\theta_{op,in} < \theta_{comf}^{ASHRAE55}} = \dfrac{\sum_{t=1}^{C}\sum_{z=1}^{Z}\left(p_{z,t}\cdot ALD_{z,t}\cdot h_t\right)}{\sum_{t=1}^{C}\sum_{z=1}^{Z}\left(p_{z,t}\cdot h_t\right)}$
	Overall	$LPD_{C}^{ASHRAE\ Adaptive}\Big|_{\theta_{op,in} \neq \theta_{comf}^{ASHRAE55}} = \dfrac{\sum_{t=1}^{C}\sum_{z=1}^{Z}\left(p_{z,t}\cdot ALD_{z,t}\cdot h_t\right)}{\sum_{t=1}^{C}\sum_{z=1}^{Z}\left(p_{z,t}\cdot h_t\right)}$
Whole year	Overheating	$LPD_{OH}^{ASHRAE\ Adaptive}\Big|_{\theta_{op,in} > \theta_{comf}^{ASHRAE55}} = \dfrac{\sum_{t=1}^{Y}\sum_{z=1}^{Z}\left(p_{z,t}\cdot ALD_{z,t}\cdot h_t\right)}{\sum_{t=1}^{Y}\sum_{z=1}^{Z}\left(p_{z,t}\cdot h_t\right)}$
	Overcooling	$LPD_{OC}^{ASHRAE\ Adaptive}\Big|_{\theta_{op,in} < \theta_{comf}^{ASHRAE55}} = \dfrac{\sum_{t=1}^{Y}\sum_{z=1}^{Z}\left(p_{z,t}\cdot ALD_{z,t}\cdot h_t\right)}{\sum_{t=1}^{Y}\sum_{z=1}^{Z}\left(p_{z,t}\cdot h_t\right)}$
	Overall	$LPD^{ASHRAE\ Adaptive}\Big|_{\theta_{op,in} \neq \theta_{comf}^{ASHRAE55}} = \dfrac{\sum_{t=1}^{Y}\sum_{z=1}^{Z}\left(p_{z,t}\cdot ALD_{z,t}\cdot h_t\right)}{\sum_{t=1}^{Y}\sum_{z=1}^{Z}\left(p_{z,t}\cdot h_t\right)}$

4.3 Likelihood of Thermal Discomfort Derived
from Ashrae Rp-884 Data

In order to derive a function that expresses the likelihood of discomfort according to the ASHRAE adaptive comfort model, a statistical analysis of the data collected in the ASHRAE RP-884 database was performed. Since the ASHRAE adaptive model is exclusively addressed to naturally ventilated buildings, only the data referred to such buildings were used.

The statistical procedure adopted to derive acceptability ranges (range of operative temperature for which PPD is smaller than a specified value) as a function of actual thermal sensation votes is similar to the one used in [6]. To keep a parallel with [6], thermal sensation votes are acronymized with ASH.

The offsets from optimal comfort temperature were calculated not only in correspondence of PPD values equal to 10 and 20 % as done in [6], but for a larger number of values of PPD, in order to derive a smooth function correlating acceptability ranges with PPD.

As a basis for the procedure, the assumptions adopted are the same used in [6]: (i) normality and equality-of-variance are applied across the database; (ii) the Fanger's relationship between PMV and PPD (Eq. 4.1) was used to relate observed thermal sensation vote (ASH) with predicted percentage of dissatisfied; (iii) a linear dependence of thermal sensation vote from operative temperature is assumed. A consequence of these assumptions is that the relationship between indoor operative temperature and PPD is symmetric with respect to the optimal comfort temperature.

The first phase of the analysis consisted in determining the statistical significance of the data collected for each building of the sample, in order to exclude those buildings characterized by few observations or rather small variations of indoor temperatures with respect to the outdoor variation from further analysis.

From the total set of 45 buildings classified as naturally ventilated, one was excluded since the thermal sensation votes (ASH) were missing. Then, the statistical significance of the remaining sample was tested using the F-test. Eight buildings failed to achieve a residual probability, p, related to F, smaller than 0.05 and were also eliminated by the set of data. A summary of the statistical analysis is reported in Table 4.5.

For each of the remaining 36 buildings, the ASHs were averaged within 0.5 °C temperature bins, and each pair of data (average ASH and mean value of the temperature bin) was weighted by the number of observations contained in each bin, in order to minimize the impact of those outliers that correspond to a low number of observations.

A linear regression was conducted for correlating the bin mean thermal sensation votes with the mean temperature values of each bin, for each building. An example of the linear regression analysis applied to the data of a building is reported in Fig. 4.2.

Table 4.5 Summary of the weighted linear regression coefficients of the bin average thermal sensation votes on indoor operative temperature

Description	Value(s)
Total number of naturally ventilated buildings of ASHRAE RP-884 database	45
Number of building with missing values	1
Number of building with regression models failing 95 % significance	8
Number of buildings with regression models achieving 95 % significance	36 (80.0 % of total buildings)
Mean model y-intercept (± standard deviation)	− 6.22 (± 3.098)
Mean model gradient (± standard deviation)	0.26 (± 0.122)

The regression models for every building were verified with respect to the graphs reported in the Appendix A of the ASHRAE RP-884 report [6].

Assuming that the observed thermal sensation votes (ASH) are distributed around their mean with the same variance of PMV, the PMV/PPD relationship can be used to calculate the two values of ASH (one positive and one negative) from a given PPD. Therefore, an array has been populated with the two values of ASH for each PPD value of the series 5, 7, 10, 20, 30, 40, 50, 60, 70, 80, 90 and 99 %. Then, for each building, using its own linear regression model (e.g., Fig. 4.2), operative temperatures were derived from each ASH value inside the array. Finally, the two operative temperatures corresponding to the same value of PPD were subtracted in order to obtain the acceptability range (in kelvin) for a given PPD. Each half-range represents the offset in kelvin from optimal comfort temperature per each value of the array. The offset in kelvin from the optimal comfort temperature calculated for each building were averaged for each PPD value obtaining the mean offset in kelvin from the optimal comfort temperature for each value of the PPD-array.

Finally, the likelihood of dissatisfied was fitted using a logistic regression over the pairs constituted by PPD values and the corresponding mean of deviations from the optimal comfort temperature, expressed in kelvin (Fig. 4.3).

The equation derived from the logistic regression analysis is called *ASHRAE Likelihood of Dissatisfied* (ALD).

$$ALD\left(\Delta\theta_{op}\right) \equiv \frac{e^{0.008 \cdot \Delta\theta_{op}^2 + 0.406 \cdot \Delta\theta_{op} - 3.050}}{1 + e^{0.008 \cdot \Delta\theta_{op}^2 + 0.406 \cdot \Delta\theta_{op} - 3.050}} \in [0.05, 1.00). \qquad (4.5)$$

where $\Delta\theta_{op} = \left|\theta_{op,in} - \theta_{comf,ASHRAE}\right|$.

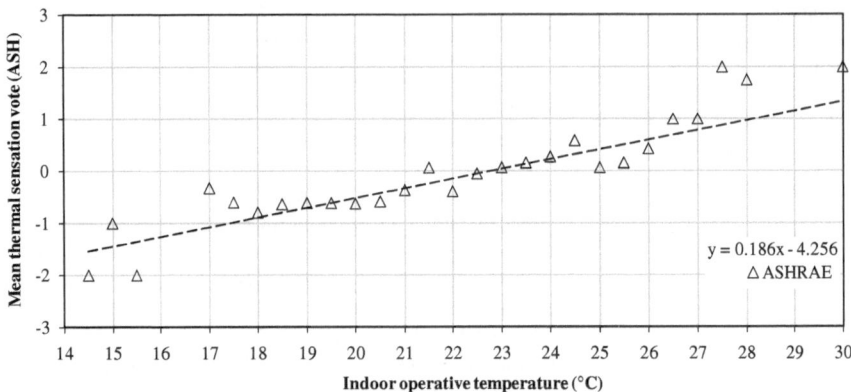

Fig. 4.2 Example of a correlation between the indoor operative temperature and the mean thermal sensation vote for one building of the ASHRAE RP-884 database: Oxford UK (summer), NV building #1

The second order term has been introduced to better fit the deviations closer to zero where the PPD shall aim at 5 %. This derives from the assumption to use the Fanger PPD/PMD function where, if PMV is zero, then PPD is 5 %.

Compared to the result of the analysis performed in the ASHRAE RP-884 report, which was limited to the two values of PPD 10 and 20 %, the analysis developed here provides the relationship PPD/temperature offset for the entire range of PPDs.

This relationship represents the discomfort likelihood function used in the ASHRAE Adaptive version of the Long-term Percentage of Dissatisfied.

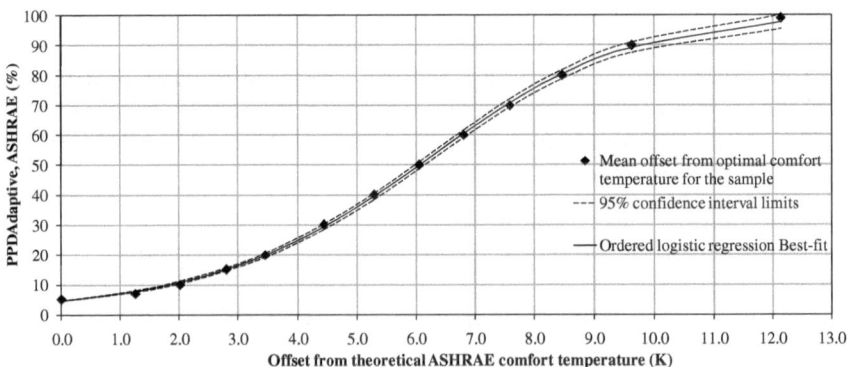

Fig. 4.3 Logistic regression ASHRAE Dissatisfied Likelihood versus the offset from comfort temperature applied to the elaboration of data from ASHRAE RP-884 database (Root mean square deviation 0.006; standard error 1.4 %)

4.4 Ranking Capability of Long-Term Percentage of Dissatisfied

In order to discuss the ranking capabilities of the new long-term discomfort index, it is calculated, in its three versions, for the 54 building variants of the large office building presented in Chap. 2 during a warm period (15 May to 30 September). The comparison of the three trends is depicted in Fig. 4.4.

The three trends are quite similar, but both the Fanger and the ASHRAE versions converge to indicate Variant 21 as the most comfortable variant among the sample whilst the EN version indicates Variant 20 as the best one.

The EN theoretical adaptive comfort temperature and the ASHRAE adaptive optimal comfort temperatures are different (Fig. 4.5) and the two versions of the index based on the adaptive comfort models are (i) very close for variants characterized by high average indoor operative temperatures, but (ii) after Variant 54, the EN version inverts the trend.

Regarding the case in analysis, the first behavior (i) is due to the different steepness of the Likelihoods of Dissatisfied. At the same value of the offset from comfort temperature, the ASHRAE function predicts a lower percentage of dissatisfied than the EN function (Fig. 4.1). Approximately, the EN and the ASHRAE versions give similar values of the likelihood of dissatisfied when the offset from ASHRAE adaptive comfort temperature is higher of 1 °C, than the offset from EN adaptive comfort temperature (Fig. 4.1). On the other hand, in this study, the ASHRAE optimal comfort temperature was calculated using the 30-day running mean of the outdoor air temperature and, for the climate in analysis – Rome –, the difference from the EN and the ASHRAE adaptive comfort temperatures is about 1 °C (Fig. 4.5). Thus, the two previous effects tend to compensate each other, and

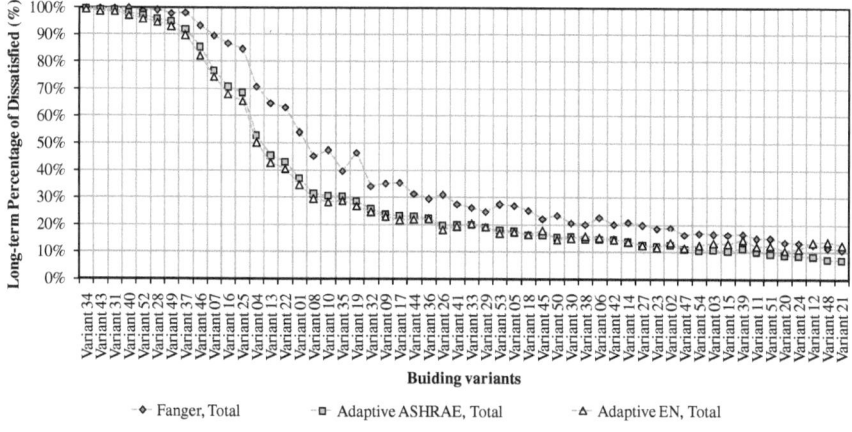

Fig. 4.4 Comparison among the three versions of the *Long-term Percentage of Dissatisfied* (it includes both overheating and overcooling)

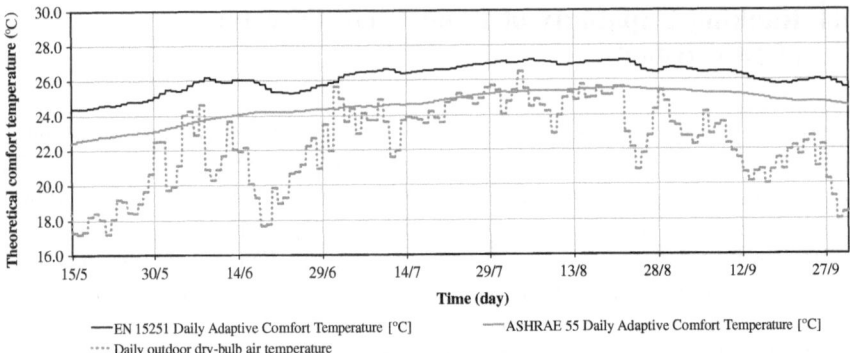

Fig. 4.5 Comparison between EN adaptive theoretical comfort temperature and ASHRAE adaptive optimal comfort temperature (according to the interpretation of outdoor running mean temperature as 30-day running mean) for a given climate (Rome)

produce similar results of likelihood of dissatisfied in the variants dominated by overheating.

The second effect (ii) is again due to the same reason: EN adaptive comfort temperature is higher than the ASHRAE one, and overcooling weighs more on the EN index. In order to detect this behavior, the indices are disaggregated in both overheating and overcooling and the sole overheating and overcooling portions are compared in Fig. 4.6. For the variants after Variant 54, the EN Long-term Percentages of Dissatisfied due to overcooling becomes larger than the one due to overheating. This explains why, in the EN version, the overall index grows for those variants with the lowest average operative temperatures. Note that, based on their definitions, the overall index does not correspond to the sum of the single overheating and overcooling indices.

The two versions based on the Fanger model and on the ASHRAE adaptive model similarly evaluate the different building variants, although the absolute values of the indices are different. This behavior is more evident for those variants characterized by very low value of the Long-term Percentage of Dissatisfied, i.e., those variants that offer a higher thermal comfort quality.

This behavior indicates that the proposed index is able to similarly assess the thermal performance of a building whether it is evaluated according to the Fanger model or whether it is evaluated according to the ASHRAE adaptive model. This is of main importance for the optimization procedure since it allows avoiding the introduction of a discontinuity if there is the necessity to change the reference comfort model, e.g., from the ASHRAE adaptive to the Fanger one.

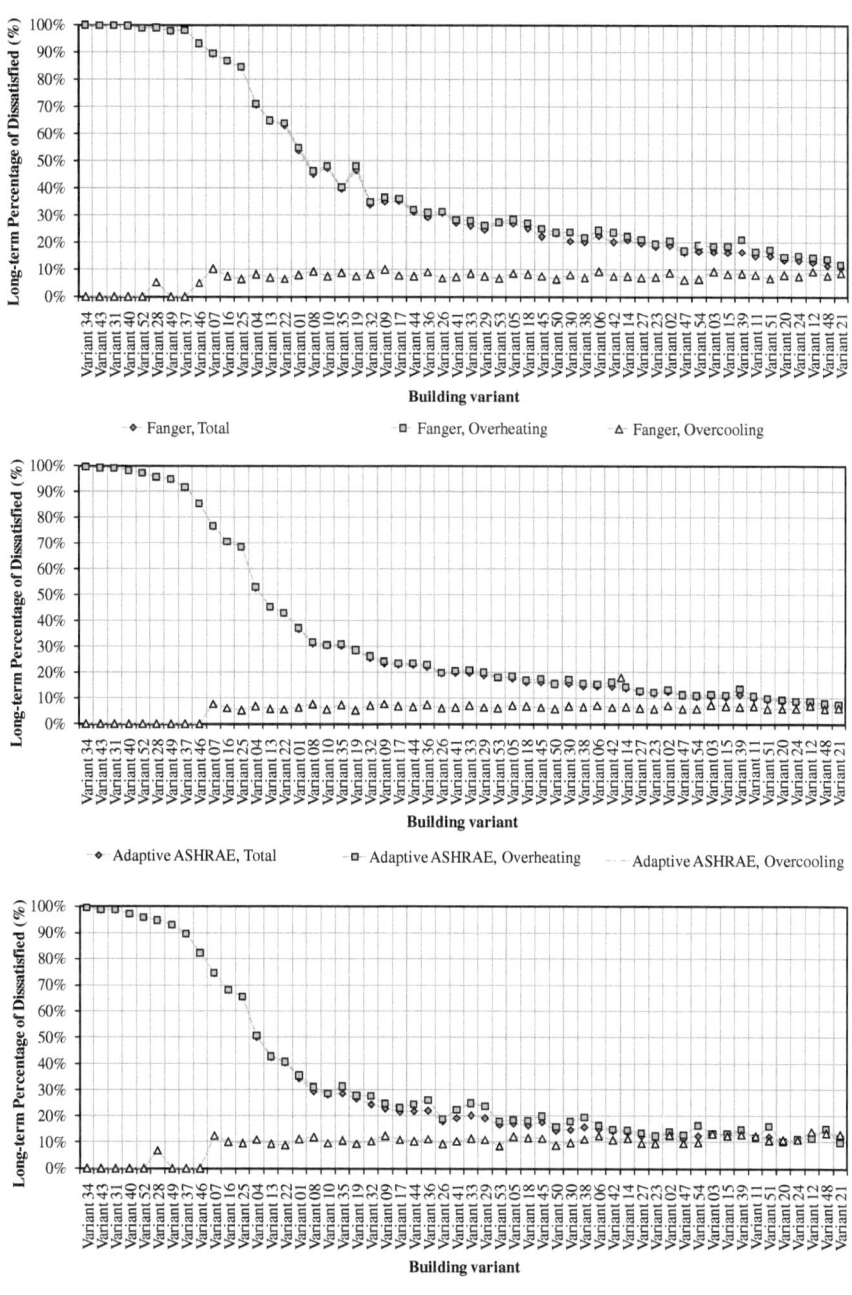

Fig. 4.6 Comparison of the index expressed for the sole overheating, the sole overcooling and both overheating and overcooling

4.5 Integrating the Proposed Index in EnergyPlus

The calculation of the indices is executed starting from the data simulated by EnergyPlus version 6.0.0.23. However, an external post-processing elaboration of thermal outputs reported by the software is required for calculating the proposed index. Therefore, in order to support an automated optimization process based on the reduction of the seasonal long-term discomfort indices, this step prevents an automated workflow. To overcome this barrier, a number of calculation algorithms have been included in EnergyPlus for calculating the three versions of the new long-term discomfort index and for reporting them as additional variables directly

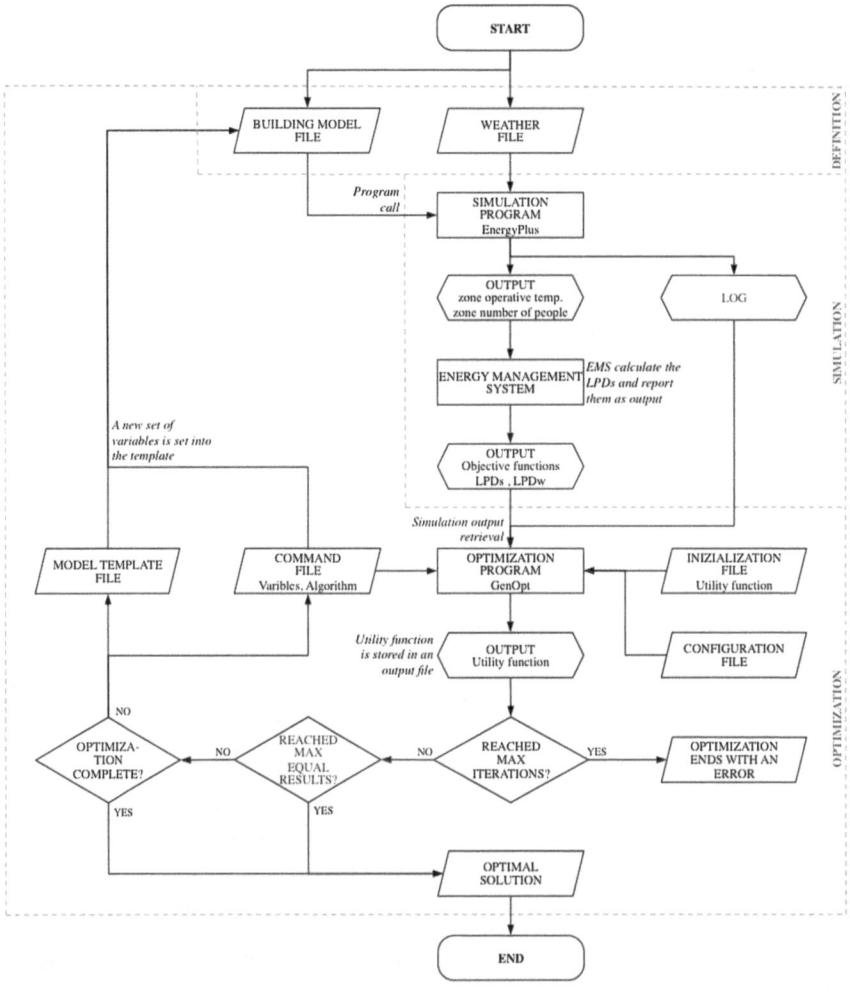

Fig. 4.7 Flow chart of the proposed comfort-optimization procedure

generated as an EnergyPlus output. In this way, the optimization algorithm can read the discomfort index from the output files of a certain generation run (say the n-ties) and take decisions in order to generate the next generation run (n + 1).

The computer codes have been written in the programming language Energy-Plus Runtime Language (ERL) and, then, integrated into EnergyPlus, through the Energy Management System module (EMS) [11, 12]. Excerpts of resulting pro-graming codes are reported in Appendix B.

A flow chart of the optimization procedure is reported in Fig. 4.7. The starting and ending days for winter, or cold period, and summer, or warm period, shall be indicated in two schedule:year called Winter and Summer. For a specified period (cold and warm), for each zone inside the building and for each time step of the simulation, EMS loads the current indoor operative temperature (a mean of dry-bulb air and mean radiant temperatures) and subtracts it from a specified theoretical daily comfort temperature (EN or ASHRAE). Then, it calculates current Likelihood of Dissatisfied, multiplies it for the current number of people inside the individual zone and stores it into a variable. At the same time, the number of people inside the zone is accumulated in another variable. The timescale of the calculation is the time step used for the simulation and each variable is averaged in the current hour. All products between Likelihood of Dissatisfied and the number of people inside the zone, and the accumulated number of people inside each zone are stored per each zone. Finally, the code calculates the Long-term Percentage of Dissatisfied for the specified period. It is calculated for the two specified seasons, and the overall value of the index (overheating plus overcooling) and the values due to the sole over-heating and the sole overcooling are reported as outputs.

4.6 Conclusions

The long-term discomfort indices accumulate (in various ways) the values of the likelihood of discomfort indices over an interval of time and over the entire building. The likelihood of discomfort provides information about the short-term discomfort predicted in a specified zone of a building according to a given comfort model.

Based on the review and comparison of a number of long-term discomfort indices, presented in Chaps. 1 and 2, in this chapter three Likelihood of Dissatisfied are selected in order to provide a continuous assessment, independent of the step changes connected to the use of comfort categories.

For the Fanger comfort model, the index that estimates the likelihood of thermal discomfort perceived by a typical occupant is the Predicted Percentage of Dissat-isfied (PPD), which is directly computable from the Predicted Mean Vote (PMV), and the calculation of both these indices is already integrated in EnergyPlus.

Regarding the EN Adaptive model, the Nicol et al.'s overheating risk (NaOR) was considered adequate to estimate the likelihood of thermal discomfort, however, although it was proposed as asymmetric index, it is symmetrically used to assess exceedance above and below comfort temperatures.

Regarding the ASHRAE adaptive model, an index for estimating the likelihood of thermal discomfort is missing in literature; thus it was necessary to build a new analytical function. It was called ASHRAE Likelihood of Dissatisfied (ALD), and has been determined via a logistic regression analysis performed on the data collected in the ASHRAE RP-884 database.

Since the calculation of NaOR and ALD are not included in EnergyPlus, programming codes have been written in the EnergyPlus Reference Language to calculate them directly in EnergyPlus.

Based on the above three Likelihood of Dissatisfied functions, a new long-term discomfort index has been proposed and it has been called *Long-term Percentage of Dissatisfied* (LPD), which is available in three versions, one for every comfort model. The proposed new index for the long-term evaluation of thermal discomfort aims at satisfying all features highlighted in Chaps. 1 and 2. Specifically, it (i) is applicable for both free-running and mechanically cooled buildings; (ii) reflects the nonlinear relationship between thermal discomfort and the exceedance from the theoretical comfort temperature for every comfort model; (iii) considers overall discomfort as related to the number of the occupants in the various zones at various time steps inside the building; (iv) is applicable to the evaluation of both summer and winter thermal discomfort; (v) can estimate thermal discomfort due to the upper and lower exceedance from the theoretical comfort temperature. Its three versions, when applied to the evaluation of the 54 variants of the test office building in Rome, provide reasonably similar trends even if the three optimal comfort targets are different. The LPDs based on adaptive models show trends that are close, but diverge if the overcooling calculated according to the EN 15251 is sensitive in a certain building variant. The LPDs based on the Fanger and the ASHRAE adaptive models show trends with a similar behavior in particular for those building variants with low values of LPD.

References

1. ISO: ISO 7730: Ergonomics of the thermal environment. Analytical determination and interpretation of thermal comfort using calculation of the PMV and PPD indices and local thermal comfort criteria, 3rd version. International Standard Organization, Geneva (2005)
2. CEN: EN 15251: Indoor environmental input parameters for design and assessment of energy performance of buildings addressing indoor air quality, thermal environment, lighting and acoustics. European Committee for Standardization, Brussels (2007)
3. ASHRAE: ANSI/ASHRAE 55: Thermal Environmental Conditions for Human Occupancy. American Society of Heating, Refrigerating and Air-Conditioning Engineers, Atlanta (2004)
4. D.F. Kalz, J. Pfafferott, Comparative evaluation of natural ventilated and mechanical cooled non-residential buildings in germany: thermal comfort in summer, in *Proceedings of Conference: Adapting to Change: New Thinking on Comfort Cumberland Lodge*, Windsor (2010)
5. P.O. Fanger, *Thermal comfort* (Danish Technical Press, Copenhagen, 1970)

6. R.J. de Dear, G.S. Brager, D. Cooper, Developing an adaptive model of thermal comfort and preference. Final report, ASHRAE RP-884 (1997)
7. J.F. Nicol, J. Hacker, B. Spires, H. Davies, Suggestion for new approach to overheating diagnostic. in *Proceedings of Conference: Air Conditioning and the Low Carbon Cooling Challenge, Cumberland Lodge*, Windsor (2008)
8. J.F. Nicol, K. McCartney, Final report of Smart Controls and Thermal Comfort (SCATs) Project. Report to the European Commission of the Smart Controls and Thermal Comfort project, Oxford Brookes University, UK (2001)
9. I. Griffiths, Thermal comfort studies in buildings with passive solar features: field studies. Report to Commission of the European Community, ENS35 909, UK (1990)
10. S. Borgeson, G.S. Brager, Comfort exceedance metrics in mixed-mode buildings. in *Proceedings of Conference: Adapting to Change: New Thinking on Comfort Cumberland Lodge*, Windsor (2010)
11. P.G. Ellis, P.A. Torcellini, D.B Crawley, Simulation of energy management systems In Energyplus. in *Proceedings of Building simulation Conference*, Beijing, 1-9 (2007)
12. US DOE: Application Guide for EMS: Energy Management System Used guide. U.S. Department of Energy (2010)

Chapter 5
Conclusions and Future Developments

In the last decade, a number of long-term discomfort indices have been proposed in the scientific literature, standards and guidelines, for describing in a concise way the long-term thermal comfort conditions in buildings. Since some of the indices are based on thermal comfort models whilst others derive from rules of thumb, since they are considerably different in their structure and significance and since a systematic collection of those was missing, a hopefully exhaustive review, a contrasting and comparing analysis and a gap analysis of the existing long-term discomfort indices have been carried out and have been presented in this work.

Since none of the analyzed indices fulfilled all requirements that an ideal long-term discomfort index should provide, a new long-term global discomfort index, called *Long-term Percentage of Dissatisfied* (LPD) has been proposed. The new index is delivered in three versions to be used with the Fanger model, the European adaptive model and the American adaptive model. Specifically, it (i) is applicable for both free-running and mechanically cooled buildings; (ii) reflects the nonlinear relationship between subjective discomfort votes and the exceedance from the theoretical comfort temperature; (iii) weighs discomfort based on zone occupancy rates rather than on a geometrical metric of the building; (iv) is applicable to the evaluation of discomfort during both summer and winter; (v) estimates possible discomfort due to the upper and lower exceedance from the theoretical comfort temperature, respectively overheating and overcooling; (vi) is independent of the abrupt boundaries of the comfort categories, which are subject of discussion in the literature.

Due to the lack in literature of a reliable likelihood of dissatisfied function based on the ASHRAE adaptive model, a new short-term local discomfort index, called *ASHRAE Likelihood of Dissatisfied* (ALD), has been also built by reanalyzing the ASHRAE RP-884 database. It allows a continuous evaluation of comfort as a function of the distance from a given comfort temperature.

Since a number of boundary conditions that sensitively affect the calculation of such indices are not explicit in standards, also a proposal for a harmonized framework for calculating the long-term discomfort indices and a proposal for the definition of their calculation period are presented.

S. Carlucci, *Thermal Comfort Assessment of Buildings*,
PoliMI SpringerBriefs, DOI: 10.1007/978-88-470-5238-3_5,
© The Author(s) 2013

In order to facilitate the calculation of LPD, a number of programming codes have been written in the programming language EnergyPlus Runtime Language, and then, they have been integrated in EnergyPlus through the Energy Management System module.

The novel index demonstrated to be effective for assessing the case study under analysis, but it should be tested in other climatic conditions and on other typologies of buildings to verify if the outlines identified in this work are still valid and are, hence, of general application, or not.

Further studies are needed to analyze a number of issues related to the use of the three comfort models:

- Occupants' thermal perception should be investigated more in detail when, during the summer, the actual operative temperature is lower than the predicted comfort temperature. Moreover, it would be useful to verify if such perception of overcooling actually exists and if it is symmetrical to overheating as it is currently implied in the most of the comfort models (but not in every long-term discomfort index).
- Although most surveys have been performed in office buildings or climate chambers with office-like conditions and type of activity, standards also extend the use of comfort models in residential buildings. The reliability of this extension should be further investigated. Another issue regarding the extension of the scope of comfort model to residential buildings consists in investigating the discomfort perception during sleeping hours.
- The ambiguity about the meteorological input of the ASHRAE adaptive model should be solved (i.e., comfort temperatures determined on a monthly or previous running 30-day basis).
- The relationships between the ASHRAE and EN adaptive comfort models should be further analyzed in order to converge possibly towards a common and unified adaptive comfort model.

Appendix A

List of stress and discomfort indices

Year	Index name	Author(s)
1905	Wet-bulb temperature	Haldane [1]
1916	Katathermometer	Hill et al. [2]
1923	Effective temperature	Houghton and Yaglou [3]
1929	Equivalent temperature	Dufton [4]
1930	Globe-thermometer temperature	Vernon [5]
1932	Corrected effective temperature	Vernon and Warner [6]
1935	Thermo integrator	Winslow et al. [7]
1937	Operative temperature	Winslow et al. [8]
1937	Humiture	Hevener [9]
1941	Standard operative temperature	Gagge [10]
1945	Thermal acceptance ratio	Ionides et al. [11]
1945	Index of physiological effect	Robinson et al. [12]
1945	Wind chill index	Siple and Passel [13]
1946	Corrected effective temperature	Bedford [14]
1947	Predicted 4-hour sweat rate	McArdel et al. [15]
1948	Resultant temperature	Missenard et al. [16]
1950	Craig index	Craig [17]
1955	Heat stress index	Belding and Hatch [18]
1957	Wet-bulb globe temperature	Yaglou and Minard [19]
1957	Oxford index	Lind and Hellon [20]
1958	Thermal strain index	Lee [21]
1959	Temperature-humidity index	Thom [22]
1959	Equatorial comfort index	Webb [23]
1960	Index of physiological strain	Hall and Polte [24]
1960	Humiture (revisited version)	Lally and Watson [25]
1961	Cumulative discomfort index	Tennenbaum et al. [26]
1962	Cumulative effective temperature	Sohar et al. [27]
1962	Index of thermal stress	Givoni [28]
1966	Heat strain index (corrected version)	McKarns and Brief [29]
1966	Prediction of heart rate	Fuller and Brouha [30]

(continued)

S. Carlucci, *Thermal Comfort Assessment of Buildings*,
PoliMI SpringerBriefs, DOI: 10.1007/978-88-470-5238-3,
© The Author(s) 2013

(continued)

Year	Index name	Author(s)
1967	Effective radiant field	Gagge et al. [31]
1970	Predicted mean vote	Fanger [32]
1970	Prescriptive zone	Lind [33]
1971	New effective temperature	Gagge et al. [34]
1971	Wet globe temperature	Botsford [35]
1971	Humid operative temperature	Nishi and Gagge [36]
1971	Apparent temperature	Steadman [37]
1972	Predicted body core temperature	Givoni and Goldman [38]
1972	Skin wettedness	Kerslake [39]
1973	Predicted heart rate	Givoni and Pandolf [40]
1974	Standard effective temperature	Gonzales et al. [41]
1978	Skin wettedness	Gonzales et al. [42]
1979	Fighter index of thermal stress	Nunneley and Stribley [43]
1979	Humidex	Masterton and Richardson [44]
1979	Apparent temperature	Steadman [45]
1980	Equivalent uniform temperature	Wray [46]
1981	Effective heat strain index	Kamon and Ryan [47]
1982	Predicted sweat loss	Shapiro et al. [48]
1982	Humisery	Weiss [49]
1982	Humiditure	Weiss [49]
1984	Munich energy balance model for individuals	Höppe [50]
1985	Skin temperature energy balance index	De Freitas [51]
1985	Heat budget index	De Freitas [51]
1986	Predicted mean vote (modified version)	Gagge et al. [52]
1987	Survival time outdoors in extreme cold	De Freitas and Symon [53]
1987	Tropical summer index	Bureau of Indian Standards [54]
1987	Summer simmer index	Pepi [55]
1988	Wind chill-equivalent temperature	Beshir and Ramsey [56]
1989	Required sweating	ISO 7933 [57]
1994	Man–environment heat exchange model	Blazejczyk [58]
1996	Cumulative heat strain index	Frank et al. [59]
1998	Physiological strain index	Moran et al. [60]
1998	Modified discomfort index	Moran et al. [61]
2000	New summer simmer index	Pepi [62]
2001	Environmental stress index	Moran et al. [63]
2002	CIBSE guide J criterion	CIBSE [64]
2005	Wet-bulb dry temperature	Wallace et al. [65]
2005	Relative humidity dry temperature	Wallace et al. [65]
2005	Percentage outside range	ISO 7730 [66]
2005	Degree-hour criterion	ISO 7730 [66]
2005	PPD-weighted criterion	ISO 7730 [66]
2005	Average PPD	ISO 7730 [66]
2005	Cumulative PPD	ISO 7730 [66]

(continued)

(continued)

Year	Index name	Author(s)
2006	CIBSE guide A criterion	CIBSE [67]
2007	Degree-hour criterion (modified version)	EN 15251 [68]
2007	Overheating risk	Robinson and Haldi [69]
2008	Overheating risk	Nicol et al. [70]
2010	Exceedance$_M$	Borgeson and Brager [71]

Appendix B

Excerpt of the ERL Computer Codes for Integrating the *LPD* in EnergyPlus

The programming code reported has been adapted for a two-zone building. Comments start with the symbol '!'. In order to understand the name of the variables the following codes are provided:

S   warm period or summer;
W   cold period or winter;
OH   overheating;
OC   overcooling;
Zt1   id for the thermal zone 1;
Zt2   id for the thermal zone 2;
TOT   total, overheating and overcooling;
DL   Likelihood of dissatisfied;
OCCUP   occupancy inside a specific zone;
DLxOCCUP   product between Likelihood of Dissatisfied and occupancy inside a specific zone;
Sum   summation over time;
DT   difference of temperature;

Bx; Exp are used for separate in simpler format the equations;
In order to run the codes, they have to be copied in the .idf file.
The variable SUMMER_SCH, WINTER_SCH, Top_comf_EN, Top_comf_ASHRAE, Top_Zn1, Top_Zn2, OCCUP_Zn1, OCCUP_Zn2, FangerPPD_Zn1, FangerPPD_Zn1 have to be defined such as EnergyManagementSystem:Sensor;
The variables Summer Overall Long-term Percentage of Dissatisfied, Summer Overheating Long-term Percentage of Dissatisfied, Summer Overcooling Long-term Percentage of Dissatisfied, Winter Overall Long-term Percentage of Dissatisfied, Winter Overheating Long-term Percentage of Dissatisfied, Winter overcooling Long-term Percentage of

S. Carlucci, *Thermal Comfort Assessment of Buildings*,   107
PoliMI SpringerBriefs, DOI: 10.1007/978-88-470-5238-3,
© The Author(s) 2013

Dissatisfied are the outputs:variables.

LPD Based on the FANGER Comfort Model

```
!-======       ALL       OBJECTS       IN       CLASS:
ENERGYMANAGEMENTSYSTEM:PROGRAMCALLINGMANAGER =======
   EnergyManagementSystem:ProgramCallingManager,
      Calc_FANGER_LPD,
      EndOfZoneTimestepBeforeZoneReporting,
      Progr_Calc_FANGER_LPD_Zn1,
      Progr_Calc_FANGER_LPD_Zn2,
      Progr_Calc_FANGER_LPDs;
      !-   ===========   ALL   OBJECTS   IN   CLASS:
ENERGYMANAGEMENTSYSTEM:PROGRAM ===========
```

EnergyManagementSystem:Program, Progr_Calc_FANGER_LPD_Zn1,
```
   !- Initialization to nullify the effect of the warm-up
days that run before the starting of the yearly simulation.
   IF i < 480,
      SET Sum_S_OH_DLxOCC_Zn1 = 0,
      SET Sum_S_OC_DLxOCC_Zn1 = 0,
      SET Sum_S_OH_OCCUP_Zn1 = 0,
      SET Sum_S_OC_OCCUP_Zn1 = 0,
      SET Sum_W_OH_DLxOCC_Zn1 = 0,
      SET Sum_W_OC_DLxOCC_Zn1 = 0,
      SET Sum_W_OH_OCCUP_Zn1 = 0,
      SET Sum_W_OC_OCCUP_Zn1 = 0,
      SET i = i + 1,
   ELSE,
      !- Cold period calculations
      IF (WINTER_SCH == 1),
         !- Calculation of the time step FANGER Likelihood of
Dissatisfied (DL)
         !- Winter overheating
         IF (Top_Zn1 - Top_comf_FANGER) >= 0,
            SET W_OH_DL_Zn1 = FangerPPD_Zn1,
            SET W_OH_DLxOCC_Zn1 = W_OH_DL_Zn1 * OCCUP_Zn1,
            SET W_OH_OCCUP_Zn1 = OCCUP_Zn1,
            SET W_OC_DL_Zn1 = 0,
            SET W_OC_DLxOCC_Zn1 = 0,
            SET W_OC_OCCUP_Zn1 = 0,
```

```
  ELSE,
  !_Winter overcooling
        SET W_OC_DL_Zn1 = FangerPPD_Zn1,
        SET W_OC_DLxOCC_Zn1 = W_OC_DL_Zn1 * OCCUP_Zn1,
        SET W_OC_OCCUP_Zn1 = OCCUP_Zn1,
        SET W_OH_DL_Zn1 = 0,
        SET W_OH_DLxOCC_Zn1 = 0,
        SET W_OH_OCCUP_Zn1 = 0,
  ENDIF,
  !- Running summations
  SET    Sum_W_OH_DLxOCC_Zn1   =   Sum_W_OH_DLxOCC_Zn1   +
W_OH_DLxOCC_Zn1,
  SET    Sum_W_OC_DLxOCC_Zn1   =   Sum_W_OC_DLxOCC_Zn1   +
W_OC_DLxOCC_Zn1,
  SET    Sum_W_OH_OCCUP_Zn1    =   Sum_W_OH_OCCUP_Zn1    +
W_OH_OCCUP_Zn1,
  SET    Sum_W_OC_OCCUP_Zn1    =   Sum_W_OC_OCCUP_Zn1    +
W_OC_OCCUP_Zn1,
  !- Warm period calculations
  ELSEIF (SUMMER_SCH == 1),
    !- Calculation of the time step FANGER Likelihood of
Dissatisfied (DL) * num. people inside the zone (OCCUP_Zn1)
    !_ Summer overheating
  IF (Top_Zn1 - Top_comf_FANGER) >= 0,
    SET S_OH_DL_Zn1 = FangerPPD_Zn1,
    SET S_OH_DLxOCC_Zn1 = S_OH_DL_Zn1 * OCCUP_Zn1,
    SET S_OH_OCCUP_Zn1 = OCCUP_Zn1,
    SET S_OC_DL_Zn1 = 0,
    SET S_OC_DLxOCC_Zn1 = 0,
    SET S_OC_OCCUP_Zn1 = 0,
  ELSE,
    !- Summer overcooling
    SET S_OC_DL_Zn1 = FangerPPD_Zn1,
    SET S_OC_DLxOCC_Zn1 = S_OC_DL_Zn1 * OCCUP_Zn1,
    SET S_OC_OCCUP_Zn1 = OCCUP_Zn1,
    SET S_OH_DL_Zn1 = 0,
    SET S_OH_DLxOCC_Zn1 = 0,
    SET S_OH_OCCUP_Zn1 = 0,
  ENDIF,
  SET    Sum_S_OH_DLxOCC_Zn1   =   Sum_S_OH_DLxOCC_Zn1   +
S_OH_DLxOCC_Zn1,
  SET    Sum_S_OC_DLxOCC_Zn1   =   Sum_S_OC_DLxOCC_Zn1   +
S_OC_DLxOCC_Zn1,
  SET    Sum_S_OH_OCCUP_Zn1    =   Sum_S_OH_OCCUP_Zn1    +
S_OH_OCCUP_Zn1,
```

```
  SET    Sum_S_OC_OCCUP_Zn1    =    Sum_S_OC_OCCUP_Zn1    +
S_OC_OCCUP_Zn1,
  !- Intermediate seasons
  ELSE,
    SET S_OH_DL_Zn1 = 0,
    SET S_OH_DLxOCC_Zn1 = 0,
    SET S_OH_OCCUP_Zn1 = 0,
    SET S_OC_DL_Zn1 = 0,
    SET S_OC_DLxOCC_Zn1 = 0,
    SET S_OC_OCCUP_Zn1 = 0,
    SET W_OH_DL_Zn1 = 0,
    SET W_OH_DLxOCC_Zn1 = 0,
    SET W_OH_OCCUP_Zn1 = 0,
    SET W_OC_DL_Zn1 = 0,
    SET W_OC_DLxOCC_Zn1 = 0,
    SET W_OC_OCCUP_Zn1 = 0,
  ENDIF,
ENDIF;
```

EnergyManagementSystem:Program, Progr_Calc_FANGER_LPD_Zn2,

```
  !- Initialization to nullify the effect of the warm-up
days that run before the starting of the yearly simulation.
  IF j < 480,
    SET Sum_S_OH_DLxOCC_Zn2 = 0,
    SET Sum_S_OC_DLxOCC_Zn2 = 0,
    SET Sum_S_OH_OCCUP_Zn2 = 0,
    SET Sum_S_OC_OCCUP_Zn2 = 0,
    SET Sum_W_OH_DLxOCC_Zn2 = 0,
    SET Sum_W_OC_DLxOCC_Zn2 = 0,
    SET Sum_W_OH_OCCUP_Zn2 = 0,
    SET Sum_W_OC_OCCUP_Zn2 = 0,
    SET j = j + 1,
  ELSE,
    !- Cold period calculations
    IF (WINTER_SCH == 1),
      !- Calculation of the time step FANGER Likelihood of
Dissatisfied (DL)
    !- Winter overheating
    IF (Top_Zn2 - Top_comf_FANGER) >= 0,
      SET W_OH_DL_Zn2 = FangerPPD_Zn2,
      SET W_OH_DLxOCC_Zn2 = W_OH_DL_Zn2 * OCCUP_Zn2,
      SET W_OH_OCCUP_Zn2 = OCCUP_Zn2,
      SET W_OC_DL_Zn2 = 0,
      SET W_OC_DLxOCC_Zn2 = 0,
      SET W_OC_OCCUP_Zn2 = 0,
```

```
     ELSE,
     !_Winter overcooling
         SET W_OC_DL_Zn2 = FangerPPD_Zn2,
         SET W_OC_DLxOCC_Zn2 = W_OC_DL_Zn2 * OCCUP_Zn2,
         SET W_OC_OCCUP_Zn2 = OCCUP_Zn2,
         SET W_OH_DL_Zn2 = 0,
         SET W_OH_DLxOCC_Zn2 = 0,
         SETCUP_Zn2 = 0,
     ENDIF,
     !- Running summations
     SET    Sum_W_OH_DLxOCC_Zn2   =   Sum_W_OH_DLxOCC_Zn2   +
     W_OH_DLxOCC_Zn2,
     SET    Sum_W_OC_DLxOCC_Zn2   =   Sum_W_OC_DLxOCC_Zn2   +
     W_OC_DLxOCC_Zn2,
     SET    Sum_W_OH_OCCUP_Zn2    =   Sum_W_OH_OCCUP_Zn2    +
     W_OH_OCCUP_Zn2,
     SET    Sum_W_OC_OCCUP_Zn2    =   Sum_W_OC_OCCUP_Zn2    +
     W_OC_OCCUP_Zn2,
     !- Warm period calculations
     ELSEIF (SUMMER_SCH == 1),
       !- Calculation of the time step FANGER Likelihood of
     Dissatisfied (DL) * num. people inside the zone (OCCUP_Zn2)
       !_ Summer overheating
       IF (Top_Zn2 - Top_comf_FANGER) >= 0,
         SET S_OH_DL_Zn2 = FangerPPD_Zn2,
         SET S_OH_DLxOCC_Zn2 = S_OH_DL_Zn2 * OCCUP_Zn2,
         SET S_OH_OCCUP_Zn2 = OCCUP_Zn2,
         SET S_OC_DL_Zn2 = 0,
         SET S_OC_DLxOCC_Zn2 = 0,
         SET S_OC_OCCUP_Zn2 = 0,
     ELSE,
         !- Summer overcooling
         SET S_OC_DL_Zn2 = FangerPPD_Zn2,
         SET S_OC_DLxOCC_Zn2 = S_OC_DL_Zn2 * OCCUP_Zn2,
         SET S_OC_OCCUP_Zn2 = OCCUP_Zn2,
         SET S_OH_DL_Zn2 = 0,
         SET S_OH_DLxOCC_Zn2 = 0,
         SET S_OH_OCCUP_Zn2 = 0,
     ENDIF,
     SET    Sum_S_OH_DLxOCC_Zn2   =   Sum_S_OH_DLxOCC_Zn2   +
     S_OH_DLxOCC_Zn2,
     SET    Sum_S_OC_DLxOCC_Zn2   =   Sum_S_OC_DLxOCC_Zn2   +
     S_OC_DLxOCC_Zn2,
     SET    Sum_S_OH_OCCUP_Zn2    =   Sum_S_OH_OCCUP_Zn2    +
     S_OH_OCCUP_Zn2,
```

```
    SET    Sum_S_OC_OCCUP_Zn2    =    Sum_S_OC_OCCUP_Zn2    +
S_OC_OCCUP_Zn2,
    !- Intermediate seasons
    ELSE,
        SET S_OH_DL_Zn2 = 0,
        SET S_OH_DLxOCC_Zn2 = 0,
        SET S_OH_OCCUP_Zn2 = 0,
        SET S_OC_DL_Zn2 = 0,
        SET S_OC_DLxOCC_Zn2 = 0,
        SET S_OC_OCCUP_Zn2 = 0,
        SET W_OH_DL_Zn2 = 0,
        SET W_OH_DLxOCC_Zn2 = 0,
        SET W_OH_OCCUP_Zn2 = 0,
        SET W_OC_DL_Zn2 = 0,
        SET W_OC_DLxOCC_Zn2 = 0,
        SET W_OC_OCCUP_Zn2 = 0,
    ENDIF,
    ENDIF;
```

EnergyManagementSystem:Program, Progr_Calc_FANGER_LPDs,

```
    !- Summation over different zones
    SET     S_OH_SumDen     =      Sum_S_OH_OCCUP_Zn1     +
Sum_S_OH_OCCUP_Zn2,
    SET     S_OC_SumDen     =      Sum_S_OC_OCCUP_Zn1     +
Sum_S_OC_OCCUP_Zn2,
    SET S_TOT_SumDen = S_OH_SumDen + S_OC_SumDen,
    SET     S_OH_SumNum     =      Sum_S_OH_DLxOCC_Zn1    +
Sum_S_OH_DLxOCC_Zn2,
    SET     S_OC_SumNum     =      Sum_S_OC_DLxOCC_Zn1    +
Sum_S_OC_DLxOCC_Zn2,
    SET S_TOT_SumNum = S_OH_SumNum + S_OC_SumNum,
    SET     W_OH_SumDen     =      Sum_W_OH_OCCUP_Zn1     +
Sum_W_OH_OCCUP_Zn2,
    SET     W_OC_SumDen     =      Sum_W_OC_OCCUP_Zn1     +
Sum_W_OC_OCCUP_Zn2,
    SET W_TOT_SumDen = W_OH_SumDen + W_OC_SumDen,
    SET     W_OH_SumNum     =      Sum_W_OH_DLxOCC_Zn1    +
Sum_W_OH_DLxOCC_Zn2,
    SET     W_OC_SumNum     =      Sum_W_OC_DLxOCC_Zn1    +
Sum_W_OC_DLxOCC_Zn2,
    SET W_TOT_SumNum = W_OH_SumNum + W_OC_SumNum,
    !- Calculation of the indices
    IF S_OH_SumDen == 0,
      SET S_OH_FANGER_LPD = 0,
```

```
ELSE,
   SET S_OH_FANGER_LPD = S_OH_SumNum / S_OH_SumDen,
ENDIF,
IF S_OC_SumDen == 0,
   SET S_OC_FANGER_LPD = 0,
ELSE,
   SET S_OC_FANGER_LPD = S_OC_SumNum / S_OC_SumDen,
ENDIF,
   SET S_TOT_FANGER_LPD = S_TOT_SumNum / S_TOT_SumDen,
IF W_OH_SumDen == 0,
   SET W_OH_FANGER_LPD = 0,
ELSE,
   SET W_OH_FANGER_LPD = W_OH_SumNum / W_OH_SumDen,
ENDIF,
IF W_OC_SumDen == 0,
   SET W_OC_FANGER_LPD = 0,
ELSE,
   SET W_OC_FANGER_LPD = W_OC_SumNum / W_OC_SumDen,
ENDIF,
SET W_TOT_FANGER_LPD = W_TOT_SumNum / W_TOT_SumDen;
```

LPD Based on the EN Adaptive Comfort Model

```
!-=========      ALL      OBJECTS      IN      CLASS:
ENERGYMANAGEMENTSYSTEM:PROGRAMCALLINGMANAGER ========
EnergyManagementSystem:ProgramCallingManager,
  Calc_EN_LPD,
  EndOfZoneTimestepBeforeZoneReporting,
  Progr_Calc_EN_LPD_Zn1,
  Progr_Calc_EN_LPD_Zn2,
  Progr_Calc_EN_LPDs;
  !-    ==========     ALL     OBJECTS     IN     CLASS:
ENERGYMANAGEMENTSYSTEM:PROGRAM ===========
```

EnergyManagementSystem:Program, Progr_Calc_EN_LPD_Zn1,
```
  !- Initialization to nullify the effect of the warm-up
days that run before the starting of the yearly simulation.
  IF i < 480,
     SET Sum_S_OH_DLxOCC_Zn1 = 0,
     SET Sum_S_OC_DLxOCC_Zn1 = 0,
     SET Sum_S_OH_OCCUP_Zn1 = 0,
     SET Sum_S_OC_OCCUP_Zn1 = 0,
     SET Sum_W_OH_DLxOCC_Zn1 = 0,
```

```
      SET Sum_W_OC_DLxOCC_Zn1 = 0,
      SET Sum_W_OH_OCCUP_Zn1 = 0,
      SET Sum_W_OC_OCCUP_Zn1 = 0,
      SET i = i + 1,
  ELSE,
    !- Cold period calculations
    IF (WINTER_SCH == 1),
      !- Calculation of the time step EN Likelihood of
Dissatisfied (DL)
      !- Winter overheating
      IF (Top_Zn1 - Top_comf_EN) >= 0,
        SET W_OH_DT_Zn1 = @Abs (Top_Zn1 - Top_comf_EN),
        SET W_OH_Bx_Zn1 = (W_OH_DT_Zn1 * 0.4734) - 2.607,
        SET W_OH_Exp_Zn1 = @Exp W_OH_Bx_Zn1,
        SET  W_OH_DL_Zn1  =  W_OH_Exp_Zn1  /  (1  +
W_OH_Exp_Zn1),
        SET W_OH_DLxOCC_Zn1 = W_OH_DL_Zn1 * OCCUP_Zn1,
        SET W_OH_OCCUP_Zn1 = OCCUP_Zn1,
        SET W_OC_DL_Zn1 = 0,
        SET W_OC_DLxOCC_Zn1 = 0,
        SET W_OC_OCCUP_Zn1 = 0,
    ELSE,
    !_Winter overcooling
        SET W_OC_DT_Zn1 = @Abs (Top_Zn1 - Top_comf_EN),
        SET W_OC_Bx_Zn1 = (W_OH_DT_Zn1 * 0.4734) - 2.607,
        SET W_OC_Exp_Zn1 = @Exp W_OC_Bx_Zn1,
        SET  W_OC_DL_Zn1  =  W_OC_Exp_Zn1  /  (1  +
W_OC_Exp_Zn1),
        SET W_OC_DLxOCC_Zn1 = W_OC_DL_Zn1 * OCCUP_Zn1,
        SET W_OC_OCCUP_Zn1 = OCCUP_Zn1,
        SET W_OH_DL_Zn1 = 0,
        SET W_OH_DLxOCC_Zn1 = 0,
        SET W_OH_OCCUP_Zn1 = 0,
    ENDIF,
    !- Running summations
    SET  Sum_W_OH_DLxOCC_Zn1  =  Sum_W_OH_DLxOCC_Zn1  +
W_OH_DLxOCC_Zn1,
    SET  Sum_W_OC_DLxOCC_Zn1  =  Sum_W_OC_DLxOCC_Zn1  +
W_OC_DLxOCC_Zn1,
    SET  Sum_W_OH_OCCUP_Zn1  =  Sum_W_OH_OCCUP_Zn1  +
W_OH_OCCUP_Zn1,
    SET  Sum_W_OC_OCCUP_Zn1  =  Sum_W_OC_OCCUP_Zn1  +
W_OC_OCCUP_Zn1,
      !- Warm period calculations
      ELSEIF (SUMMER_SCH == 1),
```

```
    !- Calculation of the time step EN Likelihood of
Dissatisfied (DL) * num. people inside the zone (OCCUP_Zn1)
    !_ Summer overheating
    IF (Top_Zn1 - Top_comf_EN) >= 0,
        SET S_OH_DT_Zn1 = @Abs (Top_Zn1 - Top_comf_EN),
        SET S_OH_Bx_Zn1 = (W_OH_DT_Zn1 * 0.4734) - 2.607,
        SET S_OH_Exp_Zn1 = @Exp S_OH_Bx_Zn1,
        SET  S_OH_DL_Zn1  =  S_OH_Exp_Zn1   /   (1   +
S_OH_Exp_Zn1),
        SET S_OH_DLxOCC_Zn1 = S_OH_DL_Zn1 * OCCUP_Zn1,
        SET S_OH_OCCUP_Zn1 = OCCUP_Zn1,
        SET S_OC_DL_Zn1 = 0,
        SET S_OC_DLxOCC_Zn1 = 0,
        SET S_OC_OCCUP_Zn1 = 0,
    ELSE,
        !- Summer overcooling
        SET S_OC_DT_Zn1 = @Abs (Top_Zn1 - Top_comf_EN),
        SET S_OC_Bx_Zn1 = (W_OH_DT_Zn1 * 0.4734) - 2.607,
        SET S_OC_Exp_Zn1 = @Exp S_OC_Bx_Zn1,
        SET  S_OC_DL_Zn1  =  S_OC_Exp_Zn1   /   (1   +
S_OC_Exp_Zn1),
        SET S_OC_DLxOCC_Zn1 = S_OC_DL_Zn1 * OCCUP_Zn1,
        SET S_OC_OCCUP_Zn1 = OCCUP_Zn1,
        SET S_OH_DL_Zn1 = 0,
        SET S_OH_DLxOCC_Zn1 = 0,
        SET S_OH_OCCUP_Zn1 = 0,
    ENDIF,
        SET Sum_S_OH_DLxOCC_Zn1 = Sum_S_OH_DLxOCC_Zn1 +
S_OH_DLxOCC_Zn1,
        SET Sum_S_OC_DLxOCC_Zn1 = Sum_S_OC_DLxOCC_Zn1 +
S_OC_DLxOCC_Zn1,
        SET Sum_S_OH_OCCUP_Zn1 = Sum_S_OH_OCCUP_Zn1 +
S_OH_OCCUP_Zn1,
        SET Sum_S_OC_OCCUP_Zn1 = Sum_S_OC_OCCUP_Zn1 +
S_OC_OCCUP_Zn1,
  !- Intermediate seasons
  ELSE,
        SET S_OH_DL_Zn1 = 0,
        SET S_OH_DLxOCC_Zn1 = 0,
        SET S_OH_OCCUP_Zn1 = 0,
        SET S_OC_DL_Zn1 = 0,
        SET S_OC_DLxOCC_Zn1 = 0,
        SET S_OC_OCCUP_Zn1 = 0,
        SET W_OH_DL_Zn1 = 0,
        SET W_OH_DLxOCC_Zn1 = 0,
```

```
      SET W_OH_OCCUP_Zn1 = 0,
      SET W_OC_DL_Zn1 = 0,
      SET W_OC_DLxOCC_Zn1 = 0,
      SET W_OC_OCCUP_Zn1 = 0,
    ENDIF,
  ENDIF;
```

EnergyManagementSystem:Program, Progr_Calc_EN_LPD_Zn2,
!- Initialization to nullify the effect of the warm-up
days that run before the starting of the yearly simulation.
```
  IF j < 480,
    SET Sum_S_OH_DLxOCC_Zn2 = 0,
    SET Sum_S_OC_DLxOCC_Zn2 = 0,
    SET Sum_S_OH_OCCUP_Zn2 = 0,
    SET Sum_S_OC_OCCUP_Zn2 = 0,
    SET Sum_W_OH_DLxOCC_Zn2 = 0,
    SET Sum_W_OC_DLxOCC_Zn2 = 0,
    SET Sum_W_OH_OCCUP_Zn2 = 0,
    SET Sum_W_OC_OCCUP_Zn2 = 0,
    SET j = j + 1,
  ELSE,
    !- Cold period calculations
    IF (WINTER_SCH == 1),
      !- Calculation of the time step EN Likelihood of
Dissatisfied (DL)
        !- Winter overheating
      IF (Top_Zn2 - Top_comf_EN) >= 0,
        SET W_OH_DT_Zn2 = @Abs (Top_Zn2 - Top_comf_EN),
        SET W_OH_Bx_Zn2 = (W_OH_DT_Zn1 * 0.4734) - 2.607,
        SET W_OH_Exp_Zn2 = @Exp W_OH_Bx_Zn2,
        SET   W_OH_DL_Zn2   =   W_OH_Exp_Zn2   /   (1   +
W_OH_Exp_Zn2),
          SET W_OH_DLxOCC_Zn2 = W_OH_DL_Zn2 * OCCUP_Zn2,
          SET W_OH_OCCUP_Zn2 = OCCUP_Zn2,
          SET W_OC_DL_Zn2 = 0,
          SET W_OC_DLxOCC_Zn2 = 0,
          SET W_OC_OCCUP_Zn2 = 0,
        ELSE,
        !_Winter overcooling
          SET W_OC_DT_Zn2 = @Abs (Top_Zn2 - Top_comf_EN),
          SET W_OC_Bx_Zn2 = (W_OH_DT_Zn1 * 0.4734) - 2.607,
          SET W_OC_Exp_Zn2 = @Exp W_OC_Bx_Zn2,
          SET   W_OC_DL_Zn2   =   W_OC_Exp_Zn2   /   (1   +
W_OC_Exp_Zn2),
```

```
        SET W_OC_DLxOCC_Zn2 = W_OC_DL_Zn2 * OCCUP_Zn2,
        SET W_OC_OCCUP_Zn2 = OCCUP_Zn2,
        SET W_OH_DL_Zn2 = 0,
        SET W_OH_DLxOCC_Zn2 = 0,
        SET W_OH_OCCUP_Zn2 = 0,
      ENDIF,
      !- Running summations
        SET  Sum_W_OH_DLxOCC_Zn2  =  Sum_W_OH_DLxOCC_Zn2  +
W_OH_DLxOCC_Zn2,
        SET  Sum_W_OC_DLxOCC_Zn2  =  Sum_W_OC_DLxOCC_Zn2  +
W_OC_DLxOCC_Zn2,
        SET  Sum_W_OH_OCCUP_Zn2  =  Sum_W_OH_OCCUP_Zn2  +
W_OH_OCCUP_Zn2,
        SET  Sum_W_OC_OCCUP_Zn2  =  Sum_W_OC_OCCUP_Zn2  +
W_OC_OCCUP_Zn2,
    !- Warm period calculations
    ELSEIF (SUMMER_SCH == 1),
      !- Calculation  of  the  time  step  EN  Likelihood  of
Dissatisfied (DL) * num. people inside the zone (OCCUP_Zn2)
      !_ Summer overheating
      IF (Top_Zn2 - Top_comf_EN) >= 0,
        SET S_OH_DT_Zn2 = @Abs (Top_Zn2 - Top_comf_EN),
        SET S_OH_Bx_Zn2 = (W_OH_DT_Zn1 * 0.4734) - 2.607,
        SET S_OH_Exp_Zn2 = @Exp S_OH_Bx_Zn2,
        SET S_OH_DL_Zn2 = S_OH_Exp_Zn2 / (1 + S_OH_Exp_Zn2),
        SET S_OH_DLxOCC_Zn2 = S_OH_DL_Zn2 * OCCUP_Zn2,
        SET S_OH_OCCUP_Zn2 = OCCUP_Zn2,
        SET S_OC_DL_Zn2 = 0,
        SET S_OC_DLxOCC_Zn2 = 0,
        SET S_OC_OCCUP_Zn2 = 0,
    ELSE,
        !- Summer overcooling
        SET S_OC_DT_Zn2 = @Abs (Top_Zn2 - Top_comf_EN),
        SET S_OC_Bx_Zn2 = (W_OH_DT_Zn1 * 0.4734) - 2.607,
        SET S_OC_Exp_Zn2 = @Exp S_OC_Bx_Zn2,
        SET S_OC_DL_Zn2 = S_OC_Exp_Zn2 / (1 + S_OC_Exp_Zn2),
        SET S_OC_DLxOCC_Zn2 = S_OC_DL_Zn2 * OCCUP_Zn2,
        SET S_OC_OCCUP_Zn2 = OCCUP_Zn2,
        SET S_OH_DL_Zn2 = 0,
        SET S_OH_DLxOCC_Zn2 = 0,
        SET S_OH_OCCUP_Zn2 = 0,
      ENDIF,
      SET  Sum_S_OH_DLxOCC_Zn2  =  Sum_S_OH_DLxOCC_Zn2  +
S_OH_DLxOCC_Zn2,
```

```
    SET   Sum_S_OC_DLxOCC_Zn2  =  Sum_S_OC_DLxOCC_Zn2  +
S_OC_DLxOCC_Zn2,
    SET   Sum_S_OH_OCCUP_Zn2  =  Sum_S_OH_OCCUP_Zn2  +
S_OH_OCCUP_Zn2,
    SET   Sum_S_OC_OCCUP_Zn2  =  Sum_S_OC_OCCUP_Zn2  +
S_OC_OCCUP_Zn2,
  !- Intermediate seasons
  ELSE,
    SET S_OH_DL_Zn2 = 0,
    SET S_OH_DLxOCC_Zn2 = 0,
    SET S_OH_OCCUP_Zn2 = 0,
    SET S_OC_DL_Zn2 = 0,
    SET S_OC_DLxOCC_Zn2 = 0,
    SET S_OC_OCCUP_Zn2 = 0,
    SET W_OH_DL_Zn2 = 0,
    SET W_OH_DLxOCC_Zn2 = 0,
    SET W_OH_OCCUP_Zn2 = 0,
    SET W_OC_DL_Zn2 = 0,
    SET W_OC_DLxOCC_Zn2 = 0,
    SET W_OC_OCCUP_Zn2 = 0,
  ENDIF,
ENDIF;
```

EnergyManagementSystem:Program, Progr_Calc_EN_LPDs,

```
  !- Summation over different zones
    SET     S_OH_SumDen     =     Sum_S_OH_OCCUP_Zn1     +
Sum_S_OH_OCCUP_Zn2,
    SET     S_OC_SumDen     =     Sum_S_OC_OCCUP_Zn1     +
Sum_S_OC_OCCUP_Zn2,
    SET S_TOT_SumDen = S_OH_SumDen + S_OC_SumDen,
    SET     S_OH_SumNum     =     Sum_S_OH_DLxOCC_Zn1     +
Sum_S_OH_DLxOCC_Zn2,
    SET     S_OC_SumNum     =     Sum_S_OC_DLxOCC_Zn1     +
Sum_S_OC_DLxOCC_Zn2,
    SET S_TOT_SumNum = S_OH_SumNum + S_OC_SumNum,
    SET     W_OH_SumDen     =     Sum_W_OH_OCCUP_Zn1     +
Sum_W_OH_OCCUP_Zn2,
    SET     W_OC_SumDen     =     Sum_W_OC_OCCUP_Zn1     +
Sum_W_OC_OCCUP_Zn2,
    SET W_TOT_SumDen = W_OH_SumDen + W_OC_SumDen,
    SET     W_OH_SumNum     =     Sum_W_OH_DLxOCC_Zn1     +
Sum_W_OH_DLxOCC_Zn2,
    SET     W_OC_SumNum     =     Sum_W_OC_DLxOCC_Zn1     +
Sum_W_OC_DLxOCC_Zn2,
    SET W_TOT_SumNum = W_OH_SumNum + W_OC_SumNum,
```

```
!- Calculation of the indices
  IF S_OH_SumDen == 0,
    SET S_OH_EN_LPD = 0,
  ELSE,
    SET S_OH_EN_LPD = S_OH_SumNum / S_OH_SumDen,
  ENDIF,
  IF S_OC_SumDen == 0,
    SET S_OC_EN_LPD = 0,
  ELSE,
    SET S_OC_EN_LPD = S_OC_SumNum / S_OC_SumDen,
  ENDIF,
    SET S_TOT_EN_LPD = S_TOT_SumNum / S_TOT_SumDen,
  IF W_OH_SumDen == 0,
    SET W_OH_EN_LPD = 0,
  ELSE,
    SET W_OH_EN_LPD = W_OH_SumNum / W_OH_SumDen,
  ENDIF,
  IF W_OC_SumDen == 0,
    SET W_OC_EN_LPD = 0,
  ELSE,
    SET W_OC_EN_LPD = W_OC_SumNum / W_OC_SumDen,
  ENDIF,
  SET W_TOT_EN_LPD = W_TOT_SumNum / W_TOT_SumDen;
```

LPD Based on the ASHRAE Adaptive Comfort Model

```
!-=========        ALL       OBJECTS      IN       CLASS:
ENERGYMANAGEMENTSYSTEM:PROGRAMCALLINGMANAGER ========
EnergyManagementSystem:ProgramCallingManager,
  Calc_ASHRAE_LPD,
  EndOfZoneTimestepBeforeZoneReporting,
  Progr_Calc_ASHRAE_LPD_Zn1,
  Progr_Calc_ASHRAE_LPD_Zn2,
  Progr_Calc_ASHRAE_LPDs;
  !-     ===========     ALL      OBJECTS      IN     CLASS:
ENERGYMANAGEMENTSYSTEM:PROGRAM ===========
```

EnergyManagementSystem:Program, Progr_Calc_ASHRAE_LPD_Zn1,
```
  !- Initialization to nullify the effect of the warm-up
days that run before the starting of the yearly simulation.
  IF i < 480,
    SET Sum_S_OH_DLxOCC_Zn1 = 0,
    SET Sum_S_OC_DLxOCC_Zn1 = 0,
```

```
    SET Sum_S_OH_OCCUP_Zn1 = 0,
    SET Sum_S_OC_OCCUP_Zn1 = 0,
    SET Sum_W_OH_DLxOCC_Zn1 = 0,
    SET Sum_W_OC_DLxOCC_Zn1 = 0,
    SET Sum_W_OH_OCCUP_Zn1 = 0,
    SET Sum_W_OC_OCCUP_Zn1 = 0,
    SET i = i + 1,
  ELSE,
    !- Cold period calculations
    IF (WINTER_SCH == 1),
      !- Calculation of the time step ASHRAE Likelihood of
Dissatisfied (DL)
      !- Winter overheating
      IF (Top_Zn1 - Top_comf_ASHRAE) >= 0,
        SET W_OH_DT_Zn1 = @Abs (Top_Zn1 - Top_comf_ASHRAE),
        SET W_OH_Bx_Zn1 = ((W_OH_DT_Zn1 ^ 2) * 0.006874) +
(W_OH_DT_Zn1 * 0.400679) - 3.0656,
        SET W_OH_Exp_Zn1 = @Exp W_OH_Bx_Zn1,
        SET  W_OH_DL_Zn1  =  W_OH_Exp_Zn1  /  (1  +
W_OH_Exp_Zn1),
        SET W_OH_DLxOCC_Zn1 = W_OH_DL_Zn1 * OCCUP_Zn1,
        SET W_OH_OCCUP_Zn1 = OCCUP_Zn1,
        SET W_OC_DL_Zn1 = 0,
        SET W_OC_DLxOCC_Zn1 = 0,
        SET W_OC_OCCUP_Zn1 = 0,
  ELSE,
  !_Winter overcooling
    SET W_OC_DT_Zn1 = @Abs (Top_Zn1 - Top_comf_ASHRAE),
    SET W_OC_Bx_Zn1 = ((W_OC_DT_Zn1 ^ 2) * 0.006874) +
(W_OC_DT_Zn1 * 0.400679) - 3.0656,
    SET W_OC_Exp_Zn1 = @Exp W_OC_Bx_Zn1,
    SET W_OC_DL_Zn1 = W_OC_Exp_Zn1 / (1 + W_OC_Exp_Zn1),
    SET W_OC_DLxOCC_Zn1 = W_OC_DL_Zn1 * OCCUP_Zn1,
    SET W_OC_OCCUP_Zn1 = OCCUP_Zn1,
    SET W_OH_DL_Zn1 = 0,
    SET W_OH_DLxOCC_Zn1 = 0,
    SET W_OH_OCCUP_Zn1 = 0,
  ENDIF,
  !- Running summations
  SET  Sum_W_OH_DLxOCC_Zn1  =  Sum_W_OH_DLxOCC_Zn1  +
W_OH_DLxOCC_Zn1,
  SET  Sum_W_OC_DLxOCC_Zn1  =  Sum_W_OC_DLxOCC_Zn1  +
W_OC_DLxOCC_Zn1,
  SET  Sum_W_OH_OCCUP_Zn1  =  Sum_W_OH_OCCUP_Zn1  +
W_OH_OCCUP_Zn1,
```

```
SET    Sum_W_OC_OCCUP_Zn1    =    Sum_W_OC_OCCUP_Zn1    +
W_OC_OCCUP_Zn1,
  !- Warm period calculations
  ELSEIF (SUMMER_SCH == 1),
    !- Calculation of the time step ASHRAE Likelihood of
Dissatisfied (DL) * num. people inside the zone (OCCUP_Zn1)
    !_ Summer overheating
    IF (Top_Zn1 - Top_comf_ASHRAE) >= 0,
      SET S_OH_DT_Zn1 = @Abs (Top_Zn1 - Top_comf_ASHRAE),
      SET S_OH_Bx_Zn1 = ((S_OH_DT_Zn1 ^ 2) * 0.006874) +
(S_OH_DT_Zn1 * 0.400679) - 3.0656,
      SET S_OH_Exp_Zn1 = @Exp S_OH_Bx_Zn1,
      SET S_OH_DL_Zn1 = S_OH_Exp_Zn1 / (1 + S_OH_Exp_Zn1),
      SET S_OH_DLxOCC_Zn1 = S_OH_DL_Zn1 * OCCUP_Zn1,
      SET S_OH_OCCUP_Zn1 = OCCUP_Zn1,
      SET S_OC_DL_Zn1 = 0,
      SET S_OC_DLxOCC_Zn1 = 0,
      SET S_OC_OCCUP_Zn1 = 0,
    ELSE,
      !- Summer overcooling
      SET S_OC_DT_Zn1 = @Abs (Top_Zn1 - Top_comf_ASHRAE),
      SET S_OC_Bx_Zn1 = ((S_OC_DT_Zn1 ^ 2) * 0.006874) +
(S_OC_DT_Zn1 * 0.400679) - 3.0656,
      SET S_OC_Exp_Zn1 = @Exp S_OC_Bx_Zn1,
      SET S_OC_DL_Zn1 = S_OC_Exp_Zn1 / (1 + S_OC_Exp_Zn1),
      SET S_OC_DLxOCC_Zn1 = S_OC_DL_Zn1 * OCCUP_Zn1,
      SET S_OC_OCCUP_Zn1 = OCCUP_Zn1,
      SET S_OH_DL_Zn1 = 0,
      SET S_OH_DLxOCC_Zn1 = 0,
      SET S_OH_OCCUP_Zn1 = 0,
    ENDIF,
    SET    Sum_S_OH_DLxOCC_Zn1    =    Sum_S_OH_DLxOCC_Zn1    +
S_OH_DLxOCC_Zn1,
    SET    Sum_S_OC_DLxOCC_Zn1    =    Sum_S_OC_DLxOCC_Zn1    +
S_OC_DLxOCC_Zn1,
    SET    Sum_S_OH_OCCUP_Zn1    =    Sum_S_OH_OCCUP_Zn1    +
S_OH_OCCUP_Zn1,
    SET    Sum_S_OC_OCCUP_Zn1    =    Sum_S_OC_OCCUP_Zn1    +
S_OC_OCCUP_Zn1,
  !- Intermediate seasons
  ELSE,
    SET S_OH_DL_Zn1 = 0,
    SET S_OH_DLxOCC_Zn1 = 0,
    SET S_OH_OCCUP_Zn1 = 0,
    SET S_OC_DL_Zn1 = 0,
```

```
    SET S_OC_DLxOCC_Zn1 = 0,
    SET S_OC_OCCUP_Zn1 = 0,
    SET W_OH_DL_Zn1 = 0,
    SET W_OH_DLxOCC_Zn1 = 0,
    SET W_OH_OCCUP_Zn1 = 0,
    SET W_OC_DL_Zn1 = 0,
    SET W_OC_DLxOCC_Zn1 = 0,
    SET W_OC_OCCUP_Zn1 = 0,
  ENDIF,
ENDIF;
```

EnergyManagementSystem:Program, Progr_Calc_ASHRAE_LPD_Zn2,

```
  !- Initialization to nullify the effect of the warm-up
days that run before the starting of the yearly simulation.
  IF j < 480,
    SET Sum_S_OH_DLxOCC_Zn2 = 0,
    SET Sum_S_OC_DLxOCC_Zn2 = 0,
    SET Sum_S_OH_OCCUP_Zn2 = 0,
    SET Sum_S_OC_OCCUP_Zn2 = 0,
    SET Sum_W_OH_DLxOCC_Zn2 = 0,
    SET Sum_W_OC_DLxOCC_Zn2 = 0,
    SET Sum_W_OH_OCCUP_Zn2 = 0,
    SET Sum_W_OC_OCCUP_Zn2 = 0,
    SET j = j + 1,
  ELSE,
  !- Cold period calculations
  IF (WINTER_SCH == 1),
    !- Calculation of the time step ASHRAE Likelihood of
Dissatisfied (DL)
    !- Winter overheating
    IF (Top_Zn2 - Top_comf_ASHRAE) >= 0,
      SET W_OH_DT_Zn2 = @Abs (Top_Zn2 - Top_comf_ASHRAE),
      SET W_OH_Bx_Zn2 = ((W_OH_DT_Zn2 ^ 2) * 0.006874) +
(W_OH_DT_Zn2 * 0.400679) - 3.0656,
      SET W_OH_Exp_Zn2 = @Exp W_OH_Bx_Zn2,
      SET W_OH_DL_Zn2 = W_OH_Exp_Zn2 / (1 + W_OH_Exp_Zn2),
      SET W_OH_DLxOCC_Zn2 = W_OH_DL_Zn2 * OCCUP_Zn2,
      SET W_OH_OCCUP_Zn2 = OCCUP_Zn2,
      SET W_OC_DL_Zn2 = 0,
      SET W_OC_DLxOCC_Zn2 = 0,
      SET W_OC_OCCUP_Zn2 = 0,
    ELSE,
    !_Winter overcooling
    SET W_OC_DT_Zn2 = @Abs (Top_Zn2 - Top_comf_ASHRAE),
```

```
      SET W_OC_Bx_Zn2 = ((W_OC_DT_Zn2 ^ 2) * 0.006874) +
 (W_OC_DT_Zn2 * 0.400679) - 3.0656,
      SET W_OC_Exp_Zn2 = @Exp W_OC_Bx_Zn2,
      SET W_OC_DL_Zn2 = W_OC_Exp_Zn2 / (1 + W_OC_Exp_Zn2),
      SET W_OC_DLxOCC_Zn2 = W_OC_DL_Zn2 * OCCUP_Zn2,
      SET W_OC_OCCUP_Zn2 = OCCUP_Zn2,
      SET W_OH_DL_Zn2 = 0,
      SET W_OH_DLxOCC_Zn2 = 0,
      SET W_OH_OCCUP_Zn2 = 0,
   ENDIF,
   !- Running summations
   SET    Sum_W_OH_DLxOCC_Zn2    =    Sum_W_OH_DLxOCC_Zn2    +
 W_OH_DLxOCC_Zn2,
   SET    Sum_W_OC_DLxOCC_Zn2    =    Sum_W_OC_DLxOCC_Zn2    +
 W_OC_DLxOCC_Zn2,
   SET    Sum_W_OH_OCCUP_Zn2     =    Sum_W_OH_OCCUP_Zn2     +
 W_OH_OCCUP_Zn2,
   SET    Sum_W_OC_OCCUP_Zn2     =    Sum_W_OC_OCCUP_Zn2     +
 W_OC_OCCUP_Zn2,
   !- Warm period calculations
   ELSEIF (SUMMER_SCH == 1),
     !- Calculation of the time step ASHRAE Likelihood of
 Dissatisfied (DL) * num. people inside the zone (OCCUP_Zn2)
     !_ Summer overheating
     IF (Top_Zn2 - Top_comf_ASHRAE) >= 0,
       SET S_OH_DT_Zn2 = @Abs (Top_Zn2 - Top_comf_ASHRAE),
       SET S_OH_Bx_Zn2 = ((S_OH_DT_Zn2 ^ 2) * 0.006874) +
 (S_OH_DT_Zn2 * 0.400679) - 3.0656,
       SET S_OH_Exp_Zn2 = @Exp S_OH_Bx_Zn2,
       SET S_OH_DL_Zn2 = S_OH_Exp_Zn2 / (1 + S_OH_Exp_Zn2),
       SET S_OH_DLxOCC_Zn2 = S_OH_DL_Zn2 * OCCUP_Zn2,
       SET S_OH_OCCUP_Zn2 = OCCUP_Zn2,
       SET S_OC_DL_Zn2 = 0,
       SET S_OC_DLxOCC_Zn2 = 0,
       SET S_OC_OCCUP_Zn2 = 0,
     ELSE,
       !- Summer overcooling
       SET S_OC_DT_Zn2 = @Abs (Top_Zn2 - Top_comf_ASHRAE),
       SET S_OC_Bx_Zn2 = ((S_OC_DT_Zn2 ^ 2) * 0.006874) +
 (S_OC_DT_Zn2 * 0.400679) - 3.0656,
       SET S_OC_Exp_Zn2 = @Exp S_OC_Bx_Zn2,
       SET S_OC_DL_Zn2 = S_OC_Exp_Zn2 / (1 + S_OC_Exp_Zn2),
       SET S_OC_DLxOCC_Zn2 = S_OC_DL_Zn2 * OCCUP_Zn2,
       SET S_OC_OCCUP_Zn2 = OCCUP_Zn2,
       SET S_OH_DL_Zn2 = 0,
```

```
      SET S_OH_DLxOCC_Zn2 = 0,
      SET S_OH_OCCUP_Zn2 = 0,
   ENDIF,
   SET   Sum_S_OH_DLxOCC_Zn2   =   Sum_S_OH_DLxOCC_Zn2   +
S_OH_DLxOCC_Zn2,
   SET   Sum_S_OC_DLxOCC_Zn2   =   Sum_S_OC_DLxOCC_Zn2   + .
S_OC_DLxOCC_Zn2,
   SET   Sum_S_OH_OCCUP_Zn2   =   Sum_S_OH_OCCUP_Zn2   +
S_OH_OCCUP_Zn2,
   SET   Sum_S_OC_OCCUP_Zn2   =   Sum_S_OC_OCCUP_Zn2   +
S_OC_OCCUP_Zn2,
  !- Intermediate seasons
  ELSE,
    SET S_OH_DL_Zn2 = 0,
    SET S_OH_DLxOCC_Zn2 = 0,
    SET S_OH_OCCUP_Zn2 = 0,
    SET S_OC_DL_Zn2 = 0,
    SET S_OC_DLxOCC_Zn2 = 0,
    SET S_OC_OCCUP_Zn2 = 0,
    SET W_OH_DL_Zn2 = 0,
    SET W_OH_DLxOCC_Zn2 = 0,
    SET W_OH_OCCUP_Zn2 = 0,
    SET W_OC_DL_Zn2 = 0,
    SET W_OC_DLxOCC_Zn2 = 0,
    SET W_OC_OCCUP_Zn2 = 0,
  ENDIF,
ENDIF;
```

EnergyManagementSystem:Program, Progr_Calc_ASHRAE_LPDs,

```
  !- Summation over different zones
    SET      S_OH_SumDen     =      Sum_S_OH_OCCUP_Zn1     +
Sum_S_OH_OCCUP_Zn2,
    SET      S_OC_SumDen     =      Sum_S_OC_OCCUP_Zn1     +
Sum_S_OC_OCCUP_Zn2,
    SET S_TOT_SumDen = S_OH_SumDen + S_OC_SumDen,
    SET      S_OH_SumNum     =      Sum_S_OH_DLxOCC_Zn1    +
Sum_S_OH_DLxOCC_Zn2,
    SET      S_OC_SumNum     =      Sum_S_OC_DLxOCC_Zn1    +
Sum_S_OC_DLxOCC_Zn2,
    SET S_TOT_SumNum = S_OH_SumNum + S_OC_SumNum,
    SET      W_OH_SumDen     =      Sum_W_OH_OCCUP_Zn1     +
Sum_W_OH_OCCUP_Zn2,
    SET      W_OC_SumDen     =      Sum_W_OC_OCCUP_Zn1     +
Sum_W_OC_OCCUP_Zn2,
    SET W_TOT_SumDen = W_OH_SumDen + W_OC_SumDen,
```

```
    SET      W_OH_SumNum      =      Sum_W_OH_DLxOCC_Zn1      +
Sum_W_OH_DLxOCC_Zn2,
    SET      W_OC_SumNum      =      Sum_W_OC_DLxOCC_Zn1      +
Sum_W_OC_DLxOCC_Zn2,
    SET W_TOT_SumNum = W_OH_SumNum + W_OC_SumNum,
  !- Calculation of the indices
  IF S_OH_SumDen == 0,
    SET S_OH_ASHRAE_LPD = 0,
  ELSE,
    SET S_OH_ASHRAE_LPD = S_OH_SumNum / S_OH_SumDen,
  ENDIF,
  IF S_OC_SumDen == 0,
    SET S_OC_ASHRAE_LPD = 0,
  ELSE,
    SET S_OC_ASHRAE_LPD = S_OC_SumNum / S_OC_SumDen,
  ENDIF,
    SET S_TOT_ASHRAE_LPD = S_TOT_SumNum / S_TOT_SumDen,
  IF W_OH_SumDen == 0,
    SET W_OH_ASHRAE_LPD = 0,
  ELSE,
    SET W_OH_ASHRAE_LPD = W_OH_SumNum / W_OH_SumDen,
  ENDIF,
  IF W_OC_SumDen == 0,
    SET W_OC_ASHRAE_LPD = 0,
  ELSE,
    SET W_OC_ASHRAE_LPD = W_OC_SumNum / W_OC_SumDen,
  ENDIF,
  SET W_TOT_ASHRAE_LPD = W_TOT_SumNum / W_TOT_SumDen;
```

Glossary

Adaptation Physiological, psychological or behavioral adjustment of building occupants to the interior thermal environment in order to avoid discomfort (EN 15251:2007).

Adaptive model A model that relates indoor design temperatures or acceptable temperature ranges to outdoor meteorological or climatological parameters (ASHRAE 55:2004).

Air speed, or Air velocity the rate of air movement at a point, without regard to direction (ASHRAE 55:2004).

Airtightness It is the resistance of the building envelope to inward or outward air leakage. Air leakage is driven by differential pressures across the building envelope. The mechanisms that create these differences in pressure are the combined effects of—stack (internal warm air rises) external wind (inducing positive and negative pressures on the envelope) and mechanical ventilation systems (ODPM 2006).

Buildings without mechanical cooling Buildings that do not have any mechanical cooling and rely on other techniques to reduce high indoor temperature during the warm season like moderately-sized windows, adequate sun shielding, use of building mass, natural ventilation, night time ventilation etc. for preventing overheating (EN 15251:2007).

Clo a unit used to express the thermal insulation provided by garments and clothing ensembles, where 1 clo = 0.155 m^2 °C/W (0.88 ft^2 h °F/Btu) (ASHRAE 55:2004).

Comfort, Thermal That condition of mind which expresses satisfaction with the thermal environment and is assessed by subjective evaluation (ASHRAE 55:2004).

S. Carlucci, *Thermal Comfort Assessment of Buildings*,
PoliMI SpringerBriefs, DOI: 10.1007/978-88-470-5238-3,
© The Author(s) 2013

Cooling season Part of the year during which (at least parts of the day and part of the building, usually summer) cooling appliances are needed to keep the indoor temperatures at specified levels. *Note* The length of the cooling season differs substantially from country to country and from region to region) (EN 15251:2007).

Environment, Thermal The characteristics of the environment that affect a person's heat loss (ASHRAE 55:2004).

Heat recovery unit (sensible and latent) Mechanical component that recover waste heat from another system and use it to replace heat that would otherwise come from a primary energy source.

Heating season Parts of the year during which (at least parts of the day and part of the building, usually winter) heating appliances are needed to keep the indoor temperatures at specified levels. Note. The length of the heating season differs substantially from country to country and from region to region) (EN 15251:2007).

Humidity, relative The ratio of the partial pressure (or density) of the water vapor in the air to the saturation pressure (or density) of water vapor at the same temperature and the same total pressure (ASHRAE 55:2004).

Insulation, Clothing The resistance to sensible heat transfer provided by a clothing ensemble. Expressed in clo units. Note: The definition of clothing insulation relates to heat transfer from the whole body and, thus, also includes the uncovered parts of the body, such as head and hands (ASHRAE 55:2004).

Irradiance It is the power of electromagnetic radiation incident on a surface per unit area. It is expressed in the International System of Units in $W\ m^{-2}$.

Mechanical cooling cooling of the indoor environment by mechanical means used to provide cooling of supply air, fan coil units, cooled surfaces, etc. Note: The definition is related to people's expectation regarding the internal temperature in warm seasons. Opening of windows during day and night time is not regarded as mechanical cooling. Any mechanical assisted ventilation (fans) is regarded as mechanical cooling (EN 15251:2007).

Median Median is described as the numerical value separating the higher half of a sample or a probability distribution, from the lower half. The median of a finite list of numbers can be found by arranging all observations from lowest value to highest value and picking the middle one. If there is an even number of observations, then there is no single middle value; the median is then usually defined to be the mean of the two middle values (Wikipedia, accessed 30/01/2012).

Met A unit used to describe the energy generated inside the body due to metabolic activity, defined as $58.2\ W/m^2$ ($18.4\ Btu\ h^{-1}ft^{-2}$), which is equal to the energy produced per unit surface area of an average person, seated at rest. The surface area of an average person is $1.8\ m^2$ ($19\ ft^2$) (ASHRAE 55:2004).

Metabolic rate The rate of transformation of chemical energy into heat and mechanical work by metabolic activities within an organism, usually expressed in terms of unit area of the total body surface. In this standard, this rate is expressed in met units (ASHRAE 55:2004).

Naturally conditioned spaces Those spaces where the thermal conditions of the space are regulated primarily by the opening and closing of windows by the occupants (ASHRAE 55:2004).

Neutrality, Thermal The indoor thermal index value corresponding with a mean vote of neutral on the thermal sensation scale

Predicted mean vote An index that predicts the mean value of the votes of a large group of persons on the seven-point thermal sensation scale (ASHRAE 55:2004).

Predicted percentage of dissatisfied An index that establishes a quantitative prediction of the percentage of thermally dissatisfied people determined from PMV (ASHRAE 55:2004).

Regression analysis It is a statistical tool for the investigation of relationships between variables.

Root mean square deviation It evaluates the differences between values predicted by a model or an estimator and the values actually observed from the modeled or estimated item.

$$RMSD = \sqrt{\frac{\sum_{i=1}^{n} [y_i - y(x_i)]^2}{n}}$$

Sensation, Thermal A conscious feeling commonly graded into the categories *cold, cool, slightly cool, neutral, slightly warm, warm,* and *hot;* it requires subjective evaluation (ASHRAE 55:2004).

Solar factor Also called Solar Heat Gain Coefficient (SHGC) or Total solar energy transmittance, is the fraction of incident irradiance that enters through a window, and includes both the directly transmitted portion and the absorbed and re-emitted portion. Solar Factor is given as a number between 0 and 1. This value can be specified in terms of the glass alone or can be referred the entire window assembly (Marinoski et al. 2007).

Standard deviation It quantifies of the variability of a data set. If the standard deviation is rather small, it indicates that the data set behave closely to the mean deviation, whether the standard deviation is high, the range of the data set is more spread

Statistical significance Statistical significance of an estimated relationship is the degree of confidence that the actual relationship is close to the estimated relationship.

Temperature, Air The temperature of the air surrounding the occupant (ASHRAE 55:2004).

Temperature, Dry-resultant The temperature recorded by a thermometer at the center of a blackened globe of 100 mm diameter (CIBSE 1999).

Temperature, Mean monthly outdoor air When used as input variable in Fig. 5.3.1 for the adaptive model, this temperature is based on the arithmetic average of the mean daily minimum and the mean daily maximum outdoor (dry-bulb) temperatures for the month in question (ASHRAE 55:2004).

Temperature, Mean radiant The uniform surface temperature of an imaginary black enclosure in which an occupant would exchange the same amount of radiant heat as in the actual non uniform space (ASHRAE 55:2004).

Temperature, Operative The uniform temperature of an imaginary black enclosure in which an occupant would exchange the same amount of heat by radiation and convection as in the actual non-uniform environment (ISO 7730:2005).

Temperature, Optimal comfort Operative temperature where a maximum number of the occupants can be expected to feel the indoor temperature acceptable. Note. For mechanical cooled building it corresponds to $PMV = 0$ (EN 15251:2007).

Temperature, Running mean Exponentially weighted running mean of the daily mean external air temperature $\theta_{\overline{d},out}$ is such a series, and is calculated from the formula (EN 15251:2007)

$$\theta_{rm} = (1-\alpha)\left(\theta_{\overline{d},out-1} + \alpha \cdot \theta_{\overline{d},out-2} + \alpha^2 \cdot \theta_{\overline{d},out-3} + \ldots + \alpha^{n-1} \cdot \theta_{\overline{d},out-n}\right) \cong$$
$$\cong \left(\theta_{\overline{d},out-1} + 0.8 \cdot \theta_{\overline{d},out-2} + 0.6 \cdot \theta_{\overline{d},out-3} + 0.5 \cdot \theta_{\overline{d},out-4} + 0.4 \cdot \theta_{\overline{d},out-5} +\right.$$
$$\left. +0.3 \cdot \theta_{\overline{d},out-6} 0.2 \cdot \theta_{\overline{d},out-7}\right)\Big/ 3.8$$

Temperature, Sol-air The temperature which, under condition of and no air motion, would cause the same heat transfer into a house as that caused by the heat exchange with the outdoor air temperature, the global solar irradiance on building surfaces and the infrared radiation with surroundings and sky (Kreider et al. 2010) $.\theta_{os} = \theta_{db,out} + \frac{\alpha I}{h_{c,r}} - \frac{\Delta q_i}{h_{c,r}}$

Thermal sensation A conscious feeling commonly graded into the categories cold, cool, slightly cool, neutral, slightly warm, warm, and hot; it requires subjective evaluation (ASHRAE 55:2004).

Transmittance, Steady-state It is the rate of energy through unit area of a component, per second and per each degree of temperature of difference across the component. Thermal transmittances shall be calculated through (i) EN ISO 6946 for most of the building components (walls, roofs and floors), (ii) EN ISO 13370 for ground floors and (iii) EN ISO 10077-1/2 for windows (glazing, frames and dividers).

Ventilation rate magnitude of outdoor air flow to a room or building either through the ventilation system or infiltration through building envelope (EN 15251:2007).

Ventilation system Combination of appliances designed to supply interior spaces with outdoor air and to extract polluted indoor air. Note. The system can consist of mechanical components (e.g. combination of air handling unit, ducts and terminal units). Ventilation system can also refer to natural ventilation systems making use of temperature differences and wind with facade grills in combination with exhaust (e.g. in corridors, toilets etc.). Both mechanical and natural ventilation can be combined with operable windows. A combination of mechanical and non-mechanical components is possible (hybrid systems) (EN 15251:2007).

Zone, Occupied The region normally occupied by people within a space, generally considered to be between the floor and 1.8 m (6 ft) above the floor and more than 1.0 m (3.3 ft) from outside walls/windows or fixed heating, ventilating, or air-conditioning equipment and 0.3 m (1 ft) from internal walls (ASHRAE 55:2004).

Zone, Thermal Part of a building throughout which the internal temperature is assumed to have negligible spatial variations (EN ISO 13786:2007).

References

1. J.S. Haldane, The influence of high air temperature. J. Hygiene **5**, 494–513 (1905)
2. L. Hill, O. Griffith, M. Flack, The measurement of the rate of heat loss at body temperature by convection, radiation and evaporation. Philos. Trans. Royal Soc. **207**(B), 183–220 (1916)
3. F.C. Houghton, C.P. Yaglou, Determining equal comfort lines. J. Am. Soc. Heat Ventilation Eng. **29**, 165–176 (1923)
4. A.F. Dufton, The eupatheostat. J. Sci. Instr. **6**, 249–251 (1929)
5. H.M. Vernon, J. Physiol. Lond. **70** (1930)
6. H.M. Vernon, C.G. Warner, The influence of the humidity of the air on capacity for work at high temperatures. J. Hygiene **32**, 431–462 (1932)
7. C.E.A. Winslow, A.P. Gagge, L. Greenburg, I.M. Moriyama, E.J. Rodee, Am. J. Hygiene **22**, 137 (1935)
8. C.E.A. Winslow, L.P. Herrington, A.P. Gagge, Physiological reactions and sensations of pleasantness under varying atmospheric conditions. Trans ASHVE **44**, 179–96 (1937)
9. O.F. Hevener, All about humiture, Weatherwise, (1937)
10. A.P. Gagge, Man, His Environment, His Comfort, Federal Proceedings (1941)

11. M. Ionides, J. Plummer, P.A. Siple, The thermal acceptance ratio. Interm report no 1, Climatology and envelope. Fed. Proc. **9**, 26 (1945)
12. S. Robinson, E.S. Turrell, S.D. Gerking, Physiologically equivalent conditions of air temperature and humidity. Am. J. Physiol. **143**, 21–32 (1945)
13. P.A. Siple, C.F. Passel, Measurement of dry atmospheric cooling in subfreezing temperatures. Proc. Am. Phil. Soc. **89**, 177–199 (1945)
14. T. Bedford, Environmental warmth and its measurement. Med Res Council Memo 17. HMSO, London (1946)
15. B. McArdle, W. Dunham, H.E. Holling, W.S.S. Ladel, J.W. Scott, M.L. Thomson, J.S. Weiner, The prediction of the physiological effects of warm and hot environments. Med Res Council, London RNP Report 47/391, London (1947)
16. A. Missenard, A thermique des ambiences: équivalences de passage, équivalences de séjours. Chaleur Indust **276**, 159–172 (1948)
17. F.N. Craig, Relation between heat balance and physiological strain in walking men clad in ventilated impermeable envelope. Fed. Proc. **9**, 26 (1950)
18. H.S. Belding, T.F. Hatch, Index for evaluating heat stress in terms of resulting physiological strain. Heat. Pip. Air. Condit. **27**, 129–136 (1955)
19. C.P. Yaglou, D. Minard, Control of heat causualties at military training centers. Am. Med. Ass. Arch. Ind. Hlth. **16**, 302–316 (1957)
20. A.R. Lind, R.F. Hallon, Assessment of physiologic severity of hot climate. J. Appl. Physiol. **11**, 35–40 (1957)
21. D.H.K. Lee, Proprioclimates of man and domestic animals. in *Climatology, Arid zone research—X, 102–25*, UNESCO, Paris (1958)
22. E.C. Thom, The discomfort index. Weatherwise **12**, 57–60 (1959)
23. C.G. Webb, An analysis of some observations of thermal comfort in an equatorial climate. Brit. J. Ind. Med. **16**(3), 297–310 (1959)
24. J.F.K. Hall, W. Polte, Physiological index of strain and body heat storage in hyperthermia. J. Appl. Physiol. **15**, 1027–1030 (1960)
25. V.E. Lally, B.F. Watson, Humiture revisited. Weatherwise **13**, 254–256 (1960)
26. J. Tennenbaum, E. Sohar, R. Adar, T. Gilat, D. Yaski, The physiological significance of the cumulative discomfort index (Cum DI). Harefuah **60**, 315–319 (1961)
27. E. Sohar, D.J. Tennenbaum, N. Robinson, A comparison of the cumulative discomfort index (Cum DI) and cumulative effective temperature (Cum ET), as obtained by meteorological data. in *Biometeorology*, ed. by S.W. Tromp, Pergamon Press, Oxford, 395–400 (1962)
28. B. Givoni, The influence of work and environmental conditions on the physiological responses and thermal equilibrium of man. in *Proceedings of UNESCO Symposium on Environmental Physiology and Psychology in Arid Conditions*, Lucknow, 199–204 (1962)
29. J.S. McKarns, R.S. Brief, Nomographs give refined estimate of heat stress index. Heat. Pip. Air. Cond. **38**, 113–136 (1966)
30. F.H. Fuller, L. Brouha, New engineering methods for evaluating the job environment. ASHRAE J. **8**, 39–52 (1966)
31. A.P. Gagge, G.M. Rapp, J.D. Hardy, Effective radiant field and operative temperature necessary for comfort with radiant heating. ASHRAE Trans. **73**, 2.1–2.9 (1967)
32. P.O. Fanger, Thermal comfort, Danish Technical Press, Copenhgen, DK (1970)
33. A.R. Lind, Effect of individual variation on upper limit of prospective zone of climates. J. Appl. Physiol. **28**, 57–62 (1970)
34. A. Gagge, A. Stolwijk, Y. Nishi, An effective temperature scale based on a simple model of human physiological regulatory response. ASHRAE Trans. **77**, 247–257 (1971)
35. J.H. Botsford, A wet globe thermometer for environmental heat measurement. Am. Ind. Hyg. Assoc. J. **32**, 1–10 (1971)
36. Y. Nishi, A.P. Gagge, Humid operative temperature. A biophysical index of thermal sensation and discomfort. J. Physiol. **63**, 365–368 (1971)
37. R.G. Steadman, Indices of wind chill of clothed persons. J. Appl. Meteor. **10**, 674–683 (1971)

38. B. Givoni, R.F. Goldman, Predicting rectal temperature response to work, environment, and clothing. J. Appl. Physiol. **32**, 812–822 (1972)
39. D.M. Kerslake, The stress of hot environment. Cambridge University Press, Cambridge (1972)
40. B. Givoni, R.R. Pandolf, Predicting heart rate response to work, environment and clothing. J. Appl. Physiol. **34**, 201–204 (1973)
41. R.R. Gonzalez, Y. Nishi, A.P. Gagge, Experimental evaluation of standard effective temperature: a new biometeorological index of man's thermal discomfort. Int. J. Biometeorol. **18**, 1–15 (1974)
42. R.R. Gonzalez, L.G. Bergulnd, A.P. Gagge, Indices of thermoregulatory strain for moderate exercise in the heat. J. Appl. Physiol. **44**, 889–899 (1978)
43. S.H. Nunneley, F. Stribley, Fighter index of thermal stress (FITS): Guidance for hot-weather aircraft operations. Aviat. Space Environ. Med. **50**, 639–642 (1979)
44. J.M. Masterton, F.A. Richardson, Humidex, a method of quantifying human discomfort due to excessive heat and humidity, CLI 1-79. Environment Canada, Atmospheric Environment Service, Downsview, Ontario (1979)
45. R.G. Steadman, The assessment of sultriness. part I: a temperature-humidity index based on human physiology and clothing science. J. Appl. Meteor. **18**, 861–873 (1979)
46. W.O. Wray, A simple procedure for assessing thermal comfort in passive solar heated buildings. Sol. Energy **25**, 327–333 (1980)
47. E. Kamon, C. Ryan, Effective heat strain index using pocket computer. Am. Ind. Hyg. Assoc. J. **42**, 611–615 (1981)
48. Y. Shapiro, K.B. Pandolf, R.F. Goldman, Predicting sweat loss response to exercise, environment and clothing. Eur. J. Appl. Physiol. Occup. Physiol. **48**, 83–96 (1982)
49. M. Weiss, The humisery and other measures of summer discomfort. Nat. Wea. Dig. **7**(2), 10–18 (1982)
50. P. Höppe, Die Energiebilanz des Menschen (dissertation). Wiss Mitt Meteorological Institute, University of München, p. 49 (1984)
51. C.R. De Freitas, Assessment of human bioclimate based on thermal response. Int. J. Biometeorol. **29,** 97–119 (1985).
52. A.P. Gagge, A.P. Fobelets, L.G. Berglund, A standard predictive index of human response to the thermal environment. ASHRAE Trans. **92**, 709–731 (1986)
53. C.R. De Freitas, L.V. Symon, A bioclimatic index of human survival times in the Antarctic. Pol. Rec. **23**, 651–659 (1987)
54. Bureau of Indian Standards. Handbook of functional requirements of buildings (other than industrial buildings). SP:41 (1987)
55. W.J. Pepi, The summer simmer index, Weatherwise **40**(3), (1987)
56. M.Y. Beshir, J.D. Ramsey, Heat stress indices: A review paper. Int. J. Ind. Ergonomics **3**, 89–102 (1988)
57. ISO 7933, *Hot environments—Analytical Determination and Interpretation of Thermal Stress Using Calculation of Required Sweat Rate* (International Organization for Standardization, Geneva, 1989)
58. K. Blazejczyk, New climatological-and-physiological model of human heat balance outdoor (MENEX) and its applications in bioclimatological studies in different scales. in *Bioclimatic Research of the Human Heat Balance*, ed. by K. Blazejczyk, B. Krawczyk (Polish Academy of Sciences, Institute of Geography and Spatial Organization, Warsaw, 1994), pp. 27–58
59. A. Frank, D. Moran, Y. Epstein, M. Belokopytov, Y. Shapiro, The estimation of heat tolerance by a new cumulative heat strain index. in *Environmental Ergonomics: Recent Progress and New Frontiers,* ed. by Y. Shapiro, D. Moran, Y. Epstein (Freund Pub House, London, 1996), pp. 194–197
60. D.S. Moran, A. Shitzer, K.B. Pandolf, A physiological strain index to evaluate heat stress. Am. J. Physiol. **275**, 129–134 (1998)

61. D.S. Moran, Y. Shapiro, Y. Epstein, W. Matthew, K.B. Pandolf, A modified discomfort index (MDI) as an alternative to the wet bulb globe temperature (WBGT). in *Environmental Ergonomics*, vol. 8, ed. by J.A. Hodgdon, J.H. Heaney, M.J. Buono (International Conference on Environmental Ergonomics, San Diego, 1998), pp. 77–80
62. W.J. Pepi, The New Summer Simmer Index. International Audience at the 80th Annual Meeting of the AMS at Long Beach, California, 2000
63. D.S. Moran, K.B. Pandolf, Y. Shapiro, Y. Heled, Y. Shani, W.T. Matthew, R.R. Gonzales, An environmental stress index (ESI) as a substitute for the wet bulb globe temperature (WBGT). J. Therm. Biol. **26**, 427–431 (2001)
64. Chartered Institution of Building Services Engineers, Guide J: Weather, solar and illuminance data. London, pp. 8.1–8.6, pp. 1–4 (2002)
65. R.F. Wallace, D. Kriebel, L. Punnett, D.H. Wegman, C.B. Wenger, J.W. Gardner, R.R. Gonzales, The effects of continuous hot weather training on risk of exertional heat illness. Med. Sci. Sports Exerc. **37**, 84–90 (2005)
66. ISO 7730: Ergonomics of the thermal environment—Analytical determination and interpretation of thermal comfort using calculation of the PMV and PPD indices and local thermal comfort criteria. 3rd version. International Organization for Standardization, Geneva (2005)
67. CIBSE: Guide A Chapter 1: Environmental Criteria for Design, Chartered Institution of Building Services Engineers, London (2006)
68. EN 15251: Indoor environmental input parameters for design and assessment of energy performance of buildings addressing indoor air quality, thermal environment, lighting and acoustics. European Committee for Standardization, Brussels (2007)
69. D. Robinson, F. Haldi, Model to predict overheating risk based on an electrical capacitor analogy. Energy Build. **40**, 1240–1245 (2007)
70. F. Nicol, J. Hacker, B. Spires, H. Davies, Suggestion for new approach to overheating diagnostics. Proceedings of Conference: Air Conditioning and the Low Carbon Cooling Challenge, Cumberland Lodge, Windsor, 2008
71. S. Borgeson, G.S. Brager, Comfort standards and variations in exceedance for mixed-mode buildings. Build. Res. Inform. **39**(2), 118–133 (2010)